510
33
Br
Ou
ma

: how
lives

D1483166

OUR DAYS ARE NUMBERED

OUR DAYS ARE NUMBERED

HOW MATHEMATICS
ORDERS OUR LIVES

JASON I. BROWN

McCLELLAND & STEWART

Library and Archives Canada Cataloguing in Publication

Brown, Jason Ira, 1961-
    Our days are numbered : how mathematics orders our lives / Jason Brown.

ISBN 978-0-7710-1696-7 (bound)

    1. Mathematics. 2. Mathematics–Social aspects. I. Title.

QA21.B76 2009   510   C2008-904237-9

We acknowledge the financial support of the Government of Canada through the Book Publishing Industry Development Program and that of the Government of Ontario through the Ontario Media Development Corporation's Ontario Book Initiative. We further acknowledge the support of the Canada Council for the Arts and the Ontario Arts Council for our publishing program.

Typeset in Garamond by M&S, Toronto
Printed and bound in Canada

McClelland & Stewart Ltd.
75 Sherbourne Street
Toronto, Ontario
M5A 2P9
www.mcclelland.com

1 2 3 4 5   13 12 11 10 09

To Sondra —

How Do I Love Thee? . . .

CONTENTS

1 TAKE A NUMBER, PLEASE!

Okay, okay, I hear you. Time to get up. I have hit the snooze button a few times, and the clock says 7:48 A.M. The clock is set ahead by 13 minutes, just to give me the illusion of sleeping in. I quickly do the mental arithmetic and decide I have to get up. Feet hit the floor, one-two, or one-zero, if I'm counting base 2. I flop back down on the bed for a few more minutes.

Last night was a bit of a sleepless one. I shouldn't have started thinking of a math problem just before going to bed. I tried to clear my mind, but it was a lost cause. I did manage to come up with some new approaches to the problem, and I'll check them out while I shower. I do a few more calculations – some calculus, some algebra, just for variety. I determine that, yep, I'm awake and ready to start the day.

2

OUR DAYS ARE NUMBERED

The world is divided into two kinds of people – those who understand mathematics and those who don't. I happen to be among the former, which has been both a blessing and a curse. When I was on the dating scene, being a mathematician was akin to having some form of communicable disease, or a sign of socially imposed celibacy. When I first met my wife, she asked what I did, and after I told her that I was a math professor, she quickly excused herself. I didn't see her again for months. Our next encounter was at a karaoke party at a recording studio. I had been a professional guitarist in my undergraduate days, and I brought my guitar to the studio. By this time I had learned my lesson, and I hid any semblance of my profession. (I even wore matching clothes and removed my usually present pocket protector.) By the end of the evening, several attractive women had slipped me their telephone numbers. "At last, some numbers I can relate to," I mused.

Today when I'm in a social situation, I am prepared. When I tell people what I do for a living, the most common response is a look of dismay, followed by "I always hated mathematics!" This statement is made with relish and without a hint of embarrassment. I don't think there is another profession out there that gets the same response. Do people state they've always hated English? Music? Lawn care? I think not.

It wasn't always this way. One hundred years ago or more, mathematics was fashionable. And it wasn't only the academics who enjoyed mathematics. History is replete with lawyers, judges, priests, and other laypeople who carried out research in mathematics, simply for the pleasure of it! Mathematics was seen as a cornerstone of a proper education. I remember when cash registers didn't display the amount of change. If the bill was for

$16.58, you could give the clerk $21.08 (with an actual dollar *bill*) and receive $4.50 back without anyone breaking a sweat. Things have changed – but not for the better. Mathematics plays a huge role in our lives. It lies hidden within the electronics we use, the banking we do, and the leisure activities we enjoy. I have found that being aware of the mathematics that underlies everyday life enhances my existence, makes it more meaningful. Embracing mathematics opens up new roads to travel, more unexplored terrain. Some may find it hard to fathom, but mathematics makes me more creative in the way I approach the world. And no matter what your ability in mathematics, it can do the same for you.

NUMBERS, NUMBERS EVERYWHERE . . .

Even basic mathematics, like arithmetic, can baffle some people, but a little knowledge can make a huge difference. Last week my wife and I had just finished lunch at a restaurant. I had the bill in hand and needed to calculate a 15% tip. I didn't want to be seen pulling out a calculator, even if I had had one on me, nor would I have wanted to take out a pen and calculate the tip by hand. So how can I appear suave *and* smart?

First some basics. A percentage can be converted to a decimal by moving the decimal point two spaces to the left. So 15%, which has a "hidden" decimal point on the right (all whole numbers, or integers, are like this), is 0.15 as a decimal. Likewise, 7% is 0.07 as a decimal (when you move the decimal two places to the left from 7, you need to add a zero).

You'll also need to know that the word "of" translates mathematically into *multiplication*. This seldom discussed fact is incredibly useful – it helps you to convert words into mathematics. So, for example, 15% of 200 translates into mathematics as 0.15×200, which is 30. Taking percentages is now a snap, using a calculator.

What if you want to do the arithmetic in your head? Is there an easy way to do the calculation? You bet. Here is what I do (it's simple enough to remember and easy enough to do mentally): Look at the bill's total, before taxes. Say it's $32.45. Taking 10% of a number is the same as multiplying by 0.10, or moving the decimal place of the number one spot to the left, therefore calculating 10% of a number is about as easy as it gets. I take 10% of the bill by moving the decimal place one spot to the left, and round to the nearest dollar (our result won't be exact, but close enough). So 10% of $32.45 is $3.245, and rounding to the nearest dollar I get $3.

Then I need to add 5% to get to the 15% tip. But 5% is half of 10%, so I need to take half of the 10% amount and add it in. Now half of $3 is pretty easy, $1.50, so I add this to the $3 to get a final tip of $4.50. I quickly plopped the money on the table and we were ready to leave. Not bad, eh?

If this is second nature, kudos to you, but if not, I suspect you'll find this simple little trick for calculating the tip useful, dare I say, even fun. It takes practice, but, hey, now you can be the one at the table of friends to grab the bill and calculate the tip; just put it back down quickly before you're on the hook for the whole enchilada.

By the way, when I told you the process, I explained not only the "how" but also the "why." I told you not only *how* to take 5% of the total from the 10% figure, but also *why* it works (you add back in half the 10% figure as 5% is half of 10%). This is one of the most important things about mathematics that people miss. Mathematics should make sense; moreover, you should *demand* that mathematics makes sense. Too often mathematics is taught religiously, where belief and sheer memorization are more important than reason.

WHY BE EXACT?

Now in our calculation for the tip, we rounded the 10% amount to the nearest dollar. Wouldn't it have been better to keep the exact number? Well, it depends – how much mental effort do you want to invest in the calculation? Often people think that an exact answer is necessary and that being unable to provide one means that you should run away from the calculation like city dwellers in a Godzilla movie. The question you should always ask is whether you need an exact answer, or whether a good approximation will do.

There are some times, to be sure, when you should be exact. My wife and I have a friend who told us that when she was growing up, her two brothers fought so much that their father bought them each a pair of real boxing gloves, to be used when they wanted to fight. One day, they were boxing and the older brother landed a punch on his younger brother's head that knocked him out cold. The older sibling did go and get his parents, but first he made sure that he counted exactly to 10 over his unconscious brother's body. (I guessed right away that the older brother is now a lawyer.)

Let's think back to the process I described for calculating the tip. How far off could you be? Well, when you round to the nearest dollar for the 10% calculation, you could be off by as much as 50 cents. So then when you calculate half of the 10% amount, you could be off by half this again – 25 cents. In total, the real tip (calculating 15% of the bill *precisely*) could at most be off by 75 cents. Are you okay with leaving a tip that might off by that much? If not, you could always play it safe by leaving an extra 75 cents. You can do the calculation in a flash, and the server will be happy. Everyone is a winner.

Estimation is a valuable skill, if only to check the reasonableness of numbers. I began forbidding the use of calculators on math exams in my classes at university when I found that students would accept whatever answer was spewed out, whether it was reasonable or not. If Student A wanted to multiply 42 by 75 and incorrectly entered 42 as 4 and got the answer 300, that would be it – the answer must be 300. But Student B, using estimation, would tell you that 42×75 is roughly 40×70, which is 2,800 – so that 300 is way off!

Train yourself (and your children) to routinely estimate calculations. It's a life skill that is worth a thousand calculators, give or take a few.

CALCULATING YOUR PULSE

I'm on the treadmill upstairs in my spare room, so excuse . . . me . . . if I'm a bit . . . out of breath. I take my pulse to measure the intensity of the workout. The calculation of my heart rate is another instance where I use mental arithmetic. I was taught that what you do is count the number of heartbeats in 10 seconds and multiply by 6 to get your beats per minute. "Per" is another word that translates well into mathematics – it converts as *division*. So a heart rate of 72 beats per minute is often written as $72 \frac{\text{beats}}{\text{minute}}$ or 72 beats /minute. We'll talk more about fractions of units in Chapter 2.

But for now, let's go back to the heart rate calculation. Suppose that during a 10-second interval you count 16 beats. You know that in one minute there are six sets of 10 seconds, so if in one set of 10 seconds you count 16 beats, you should have six sets of the 16 beats in 60 seconds. That is, in 60 seconds, there should be 16+16+16+16+16+16 beats, which is 16×6, or 96 beats. Your heart rate is 96 beats per minute, which is a little high, but hey, you're reading an exciting book on mathematics.

Wouldn't it be easier to count the number of beats in six seconds, and multiply by 10 to get the beats per minute? Yep, it would. The reason the rule of thumb (or of two fingers, as that's what you use to take your pulse) uses 10 seconds rather than six seconds is a consideration of accuracy. Ideally, you should start your count right after a beat, counting the next beat as 1. But when you start your count, you might be off by a beat at the beginning of your count. You might start timing right on the beat, or somewhere in the middle of a beat. No matter how careful you are, at the beginning you might still be off by a beat in your count. Also, it's possible you might be off by a fraction of a second in your timing at the end; this may add an error of another beat or so. So you could actually be off by two beats in your count. This translates into being off by 6×2=12 beats per minute in your calculation if you did your count over 10 seconds, and 10×2=20 beats per minute if you did your count over six seconds instead. Being off by 12 beats per minute in measuring your pulse is not as big a deal as being off by 20 beats.

As a shortcut to multiplying by 6 in your head, what you can do is break it up into multiplying by 5 and adding the original amount. I find it even quicker to multiply by 10 (that is, add a zero) and divide in half instead of multiplying by 5. So if I count 16 beats, I multiply by 10 to get 160. I divide that in half to get 80, then I add in the original number 16 to get 96, the number of beats per minute.

CASTING OUT NINES

There is a trick that can help you find errors in addition, subtraction, and multiplication, without resorting to using a calculator. It's called *casting out nines* and used to be regularly taught in high schools. I remember learning about casting out nines when I started reading math books for recreation. Now that may strike you as a little weird, but somehow school had wrung out every bit of the fun in mathematics, so I had to get my excitement elsewhere. I'm not sure when exactly I found out I had a natural inclination for things mathematical. I do remember having to do a math project in grade 4, and deciding to take my brother's grade 13 algebra textbook and teach myself about matrices. It didn't seem too hard. Nobody told me I couldn't or shouldn't do that. It wasn't long afterward that I found math books at the library with all sorts of interesting, mysterious things in them, things that were never taught in school.

Suppose you want to check a calculation that you've carried out, perhaps for adding a down payment and some lease payments for a car, such as $3,468+(24×$437). The $3,468 is the down payment on the vehicle, with 24 monthly payments of $437 to follow. You've done the calculation and your result is

$13,966. Is there a quick way to check your answer, other than redoing the calculation? There is, and here is all you do. You can "cast out nines," replacing a number by the sum of its digits. You can do this as often as you like until the final answer is between 0 and 8. Do this both within the calculation and for the given answer. If the final numbers disagree, you can be sure that you've made an error somewhere. In our example, $3,468+(24×$437), you can replace

- 3,468 by 3+4+6+8, which is 21, which we can replace by 2+1=3,
- 24 by 2+4=6, and
- 437 by 4+3+7, which is 14, which we again replace by 1+4=5.

Thus we replace our original calculation 3,468+(24×$437) by 3+(6×5). This becomes 3+30, or 33, and we can again add up the digits to get 3+3=6, and stop.

Meanwhile, if we add the digits of the purported answer of 13,966 we get 1+3+9+6+6=25, and again adding up the digits we get 7. Since the two numbers are different, we can conclude that we must have made a mistake somewhere.

By the way, the correct answer is in fact 13,956. If you add up its digits, you get 1+3+9+5+6=24, and 2+4=6. In this case, the two numbers match up, as they must if the answer is correct.

A couple of things come to mind. First, the procedure won't catch some errors, like transpositions of digits, such as writing 13,965 instead of 13,956, because shuffling the order of the digits doesn't affect the sum of the digits. Secondly, the process works wonderfully quickly and easily, but *why* does it work? The reasons are complicated, but I can tell you that the validity of the procedure relies on what I consider to be an amazing fact: For any whole number, the number 9 always divides the difference between the number and the sum of its digits ("divides" for two integers means "evenly," so that there is no remainder). Let's try it out. If we look at 24, 9 really does divide the difference between 24 and the sum of its digits, 6, since 24–6=18 and 9 divides 18 evenly. But maybe we got lucky! Let's try it with 13,956. The sum of the digits is 24. Does 9 divide 13,956–24=13,932? Yes, in fact 13,962=9×1,548, so 9 does indeed divide 13,932.

Now, you may say this looks like much more than coincidence, and rightly so. Either you believe it or you can demand more evidence – what mathematicians call a *proof*. I am going to ignore the temptation to provide a proof here; I'll talk about proofs in more detail later on. For now, pretend I'm a politician and just put your trust in me.

The "amazing fact" also gives a quick way to check whether 9 divides a given whole number; all you need to do is keep adding the digits, repeating if necessary, until you get a number between 1 and 9. If you get 9, 9 does indeed divide the number evenly, otherwise it doesn't. So does 9 divide 8,423,766? All we need to do is to add the digits 8+4+2+3+7+6+6, to get 36. We again add the digits, 3+6, to get 9. We conclude that indeed 9 divides 8,423,766 evenly. On the other hand, 9 does not divide, say, 7,422 evenly since the sum of the digits is 7+4+2+2, which is 15, and the sum of the digits of 15 is 6.

Checking whether 9 divides a whole number is a bit of a mathematical trick. Such a trick belongs to an area of mathematics known as *number theory*, perhaps the part of mathematics that has garnered the most attention from non-professionals, simply for its beauty. The mysterious and surprising patterns numbers (especially prime numbers) form are captivating. It is a beauty that is in the eye of the beholder, but many over the years have fallen for its allure.

A PARLOUR TRICK (EVEN IF YOU DON'T HAVE A PARLOUR)

Mastering divisibility tricks might impress your friends – but then again it might not. Here is a more mystifying trick that will leave them guessing. Take a set of dominos; each of the 28 tiles shows two numbers represented by dots, ranging from 0 to 6. Place the dominos face down on the table and ask a friend to pick any one domino, while you turn away. With your back still turned, ask her to do the following:

Step 1. Multiply one of the numbers on the domino by 5.

Step 2. Add 9 to the result.

Step 3. Multiply this number by 2.

Step 4. Add the other number on the domino to this number.

Step 5. Tell you the answer she gets.

Subtract 18 from the number she gives you. Amazingly, the two digits of the result will be the numbers on the domino (if the two-digit number is smaller than 10, treat the missing digit as zero).

Let's try it. Suppose the person pulls the following domino:

If she picks the number 3, she first multiplies 3 by 5 to get 15. Then she adds 9 to get 24, and multiplies by 2 to get 48. Finally, she adds the other domino number, 5, to get 53, which is the number she tells you. If you now subtract 18 from 53 you get 35. You can pronounce the two numbers on her domino are indeed 3 and 5!

If, on the other hand, your friend chooses the other number, 5, she would multiply 5 by 5 to get 25, add 9 to get 34, and double it to get 68. Finally, she would add in 3, the other domino number, and tell you she got 71. After hearing this number, you would subtract 18 to get 53 (5 and 3) and announce correctly that the two numbers on the domino were 5 and 3!

You can try it out a few more times, but of course that won't convince you the trick always works, or *why* it works. What you need is an argument, and a convincing one at that, to show that the trick will give the correct answer every time. (I use the word *argument* instead of *proof*, because I find the latter raises most people's blood pressure.)

Here is the argument. Suppose the two numbers on the domino are x and y (one key step in many mathematical proofs is to label unknowns with letters). If x is chosen, you multiply x by 5, to get $5x$. Then you add 9, to get $5x+9$. You multiply the result by 2, to get $2\times(5x+9)$. You need brackets because you are multiplying by 2 the *whole* quantity of $5x+9$. Now you add in the other number, y, and get $2\times(5x+9)+y$. You can use a bit of algebra to rewrite this calculation as $2\times(5x+9)+y=2\times5x+2\times9+y$ $=10x+18+y$, which you can rewrite as $10x+y+18$. This is the number you will be told (as per step 5, above). If you subtract 18, you get $10x+y+18-18$ or $10x+y$. That is, the tens digit is x, one number on the domino, and the ones digit is y, the other number on the domino. This *proves* the procedure always works.

The proof, in fact, tells you even more. First, the tens digit of the number you calculate in the end is the number on the domino chosen by the participant in step 1 (both are what we called x). Secondly, the proof works as long as x and y are at most 9, that is, as long as they are both single digits. The trick would even work with a domino set where the numbers are from 0 to 9 instead of 0 to 6 (if dominos are not available, simply ask the participant to think of two numbers from 0 to 9). Proofs can offer so much more than just a convincing argument; they can show the limits and extensions of what is true.

Such tricks do circulate on the Internet from time to time. The following is an e-mail a friend forwarded to me in 2007:

Your Age by Eating Out

1. First, pick a number of times a week that you would like to go out to eat (more than once but less than 10).
2. Multiply this number by 2.
3. Add 5.
4. Multiply by 50.
5. If you have already had your birthday this year, add 1,757; if you haven't, add 1,756.
6. Now subtract the four-digit year that you were born. You should have a three-digit number.

The first digit of this three-digit number will be your original number (the number of times you want to go out to a restaurant per week). The next two numbers will be YOUR AGE!

THIS IS THE ONLY YEAR (2007) IT WILL EVER WORK, SO SPREAD IT AROUND WHILE IT LASTS.

First, I am always leery of words capitalized for emphasis – what are they hiding? I am also wary of calculations that are passed off as some kind of magic – it ain't possible. So I sat down for a few minutes and tried to figure out what was going on, and it was really straightforward. Suppose we call the single-digit number of times you'd like to go out to eat per week x. (The mathematical "we" is similar to the "we" nurses use – minus the rectal thermometer.) Now if we carry out steps 2 through 4 we get, in turn, the numbers $2x$ (multiply by 2), $2x+5$ (add 5), and $50(2x+5)$ (multiply by 50). Simple algebra turns $50(2x+5)$ into $100x+250$.

If you've had your birthday already, then we add 1,757, to get $100x+250+1,757$, which is $100x+2007$ (aha, I recognize the current year in this formula). Now if we subtract your year of birth (let's call that y), we get $100x+(2007-y)$. As you've already had your birthday this year, $2007-y$, the difference in years since your birthday, just counts how old you are in 2007, and of course the first digit of $100x+(2007-y)$ is x, which is the number of times you wanted to go out to eat. Presto!

The argument for the case of not having your birthday yet is pretty much the same, except for the fact that if you haven't yet had your birthday in 2007, then your age is the difference between your year of birth and the previous year, 2006 (this is why you need to add 1,756 instead of 1,757 in step 5).

This reasoning explains away the mystery and so much more. First, the claim that 2007 is the only year for which the trick works is a little misleading. To make it work for 2008, just add 1 to 1,757 and 1,756; for 2009, add 2 to 1,757 and 1,756; and so on. Secondly, the proof shows that the procedure won't always work! The problem is that $2007-y$ (or $2006-y$ in the case of your birthday in 2007 not having occurred yet) needs to be a two-digit number, so that the first digit only depends on the $100x$. So, if your age is 100 or more, there's a problem. For example, if your age is now 111 and you want to go out to eat three times a week (you certainly deserve it!), then what you end the calculations with is $100\times3+111=411$, and the conclusion is that you are 11 years old and want to go out to eat four times per week. I assume that whoever made up the trick didn't bother to think about centenarians.

In just a few examples we have seen that mathematics can be very useful (in calculating tips in our head), entertaining (the domino trick), and in bridging the gaps between the two (casting out nines). In the chapters that follow we'll taste a few more mathematical treats. Take a number, please! Next!

I open the front door, pick up the paper, and head into the kitchen. I still can't shake this math problem, which is a problem as I try to navigate the random arrangement of mini hockey sticks and balls that my sons have left lying around. There's nothing like tripping and crashing into a wall to empty my mind, if only for a moment.

Anyway, it's a lovely sunny day, −2 degrees Celsius, which is 28.4 degrees Fahrenheit. And even better, I've started the day off with mathematics, not that I could help myself. I learned long ago that I can't separate the mathematical part of my life from the rest of my life. Nor would I want to.

CONVERSION CLASSES FOR THE MASSES

In the kitchen my sons are waiting impatiently for the chef (i.e., me) to prepare breakfast. I like to think of myself as handy, even in the kitchen. I'm not against trying my hand at cooking up some culinary delight – it's just that other people are.

For most families, the threat for misbehaving kids is "Wait until your father gets home!" For my sons it is "Wait until your father gets home. I'll have him make dinner. We'll see who's sorry then!"

Even though it has been a couple of months since my boys returned from their first overnight summer camp, they continue to regale me with their adventures. My eldest has told me that he was his cabin's best at "Fear Factor," where the goal was to eat whatever concoction the counsellors made up, without gagging. I guess all my recipes have paid off.

One night last week we had guests over for dinner. My wife, a nurse, headed off to work early that morning as usual, for a long, 12-hour day. I was the chef that evening, and in addition to making salmon, I made my black bean soup as well. I have a recipe for the soup that serves four that I had to convert to a recipe for seven, but no problem. I did the calculations in my head. I might not be able to cook well, but at the very least I could get the math straight.

We measure everything. From ingredients in recipes to gasoline in our tanks to our heights and weights – everything is counted, sorted, and compared. Without numbers we'd be unable to know whether we are keeping up with the Joneses. But in many situations, the units of measurement we are trying to compare may differ. In Canada, unlike in the United States, all of our measurements are officially done using the metric system – millilitres instead of ounces, litres instead of gallons, centimetres instead of inches, metres instead of yards, kilometres instead of miles, kilograms instead of pounds – the list goes on and on. It wasn't always so. The change happened when I was in my teen years, and it took me the longest time to think in metric. I still sometimes convert from Celsius to Fahrenheit when I hear a weather report. The temperatures in Celsius just aren't as meaningful to me.

All of this brings to mind the process of conversion. I am not talking about religion, but I will be proselytizing. Learning to convert from one unit to another comes up so often in everyday life that understanding how to convert with confidence is invaluable. So get ready to join your first conversion class.

A good part of what I do in mathematics is teaching – it's one of the reasons I chose it as a profession. To see the spark that occurs when mathematics comes alive to a student, when math means something more than mere symbols, is exhilarating. Unfortunately, I have found that far too often mathematics at the elementary and high school levels is taught by teachers who not only dislike mathematics but also fear it! I remember going out on a date with an elementary school teacher who told me that she taught grade 3 math but was worried about next year when she would have to teach grade 4 math. She didn't know if she would be able to handle it. Cheque, please! And I'll calculate the tip!

PROPORTIONALITY

Most (but not all) of the various units we use to measure the same quality of an item are *proportional* – doubling one doubles the other, tripling one triples the other, halving one halves the other, and so on. Weight in pounds and kilograms is like that; doubling the weight in pounds doubles the weight in kilograms. Units of length (inches, feet, yards, miles, centimetres, metres, and kilometres) are proportional as well.

For units that are proportional, you can form what is called a *conversion factor*, a fraction (yes, one of those!) that will allow you to convert from one unit to the other. All you need to memorize (or look up) is how many units of one is equal to a certain amount of the other. For example, one inch is about 2.5 centimetres; this fact gives us two conversion factors: $\frac{1 \text{ in}}{2.5 \text{ cm}}$ and $\frac{2.5 \text{ cm}}{1 \text{ in}}$.

Converting from one unit to the other is easy with conversion factors: all you need to do is to multiply the number that you want to convert by the appropriate conversion factor. So if you want to convert 11 inches into centimetres, you need to multiply by either $\frac{1\text{ in}}{2.5\text{ cm}}$ or $\frac{2.5\text{ cm}}{1\text{ in}}$. But which conversion factor do you choose?

The answer is *the one so that the units cancel properly*. So if we want to convert 11 inches, we have inches in the top (as a number by itself is the same as a fraction with that number on the top and 1 in the bottom), and we want our final answer to have centimetres (on the top). (The top of a fraction is usually called the *numerator* and the bottom the *denominator*, but when I teach, I like to use colloquial terms, terms that have more meaning.)

We need to cancel out the inches, so we need to use the conversion fraction with inches in the bottom. Something on the top of such a fraction cancels with the same item on the bottom when you multiply. So the calculation we do is

$$11\text{ in} = 11\text{ in} \times \frac{2.5\text{ cm}}{1\text{ in}} = \frac{11 \times 2.5}{1}\text{ cm} = 27.5\text{ cm}.$$

I make sure that I write down the units as well as the numbers and physically cancel the units; this cancelling ensures that I multiply by the correct conversion factor.

I can't tell you how many people struggle with conversions because they know they need to multiply or divide but can't remember which! If I am converting from inches to centimetres, do I multiply by 2.5 or divide by 2.5? The answer is easy if you write down the units. This may seem pedantic, but it's like learning to swim – at first you may need a life jacket or water wings, but eventually you can swim without these aids. (I am a lousy swimmer, but you get the point.)

GOING THE EXTRA 1,609,344 MILLIMETRES

My wife and I are in the midst of buying a new car; it's become difficult trying to arrange our lives around one vehicle. We have a large collection of brochures gathered from visits to dealerships, and as I look through one I have to laugh. One of the key dimensions for us is the length of the vehicle; our minivan and the new car will have to be parked one behind the other in our driveway.

Here is the kicker – the length of the car we are interested in is listed as being 4,800 millimetres. Millimetres! Who on earth measures a car in millimetres? For most people this measurement is probably useless. Of course, what you should do is convert the measurement into metres. Since there are 1,000 millimetres in a metre, 4,800 millimetres is 4.8 metres, or, using conversion factors:

$$4{,}800 \text{ mm} = 4{,}800 \text{ mm} \times \frac{1 \text{ m}}{1{,}000 \text{ mm}} = \frac{4{,}800 \times 1}{1{,}000} \text{ m} = 4.8 \text{ m}.$$

That's much better! A number that is reasonable. After measuring my driveway (in metres, not millimetres), I can be sure that both the new car and my old van will fit.

Now any time I take my car in for service, I'll be sure to tell the driver of the courtesy van that gives me a lift to my office appropriate directions: "Go about 1,700,000 millimetres down Robie Street . . ."

THAT LONG DISTANCE FEELING

Occasionally, when I suffer from amnesia and forget about the last long road trip we took as a family, I get the urge to climb into the van and drive as far as the eye can see. I'm nearsighted, so that only takes me down to the end of the street, but at other times we head off in search of adventure. My wife and I are notorious for getting lost. At such times I say, "It doesn't matter as long as we're all together," and then my wife starts muttering to herself. When driving in the United States, I need to convert miles to kilometres since our Canadian-made van's odometer shows only kilometres travelled. The speedometer (which shows speeds in both kilometres per hour and miles per hour) indicates that 50 miles per hour is the same as 80 kilometres per hour. As they are both "per hour," it follows that 50 miles is approximately 80 kilometres. This information is enough to form the conversion factors $\frac{50 \text{ mi}}{80 \text{ km}}$ and $\frac{80 \text{ km}}{50 \text{ mi}}$. These are the same (by cancelling out tens in the top and bottom) as $\frac{5 \text{ mi}}{8 \text{ km}}$ and $\frac{8 \text{ km}}{5 \text{ mi}}$. The fact that I want to convert *from* miles *to* kilometres means that I want to use the conversion factor with miles in the bottom (so that the units of miles will cancel out). If a road sign tells me that it's 20 miles until my exit, then I can set my trip odometer to

$$20 \text{ mi} = 20 \ \cancel{\text{mi}} \times \frac{8 \text{ km}}{5 \ \cancel{\text{mi}}} = \frac{20 \times 8}{5} \text{ km} = 32 \text{ km}.$$

I could use a calculator for the arithmetic, but in my head I can either divide 5 into 20 to get 4 and then multiply by 8 (to get 32), or I can divide the number I get in the top (160) by 10 (to get 16) and then multiply by 2 (to get 32 again), as dividing by 5 is the same as dividing by 10 (twice as much as I want to divide by, but a simple calculation) and then doubling the answer.

CONVERT . . . YOUR LIFE MAY DEPEND ON IT!

All this talk about proportionality and conversion factors may seem a bit abstruse, but in 1983 an incorrect conversion factor nearly caused a huge air disaster. Air Canada Flight 143, a Boeing 767, was flying from Montreal to Edmonton. Its fuel gauges were not working properly, so the maintenance crew calculated how much fuel the plane needed for the flight, using a manual "drip" reading that gauged the depth of the fuel in the tanks. The specific gravity of jet fuel – 1.77 *pounds* per litre – is needed to make the proper drip calculations. But this new plane used metric (Canada started metrication in 1970 but fuel sales in metric only began in 1981), and the crew should have used the specific gravity factor in *kilograms* per litre. Since there are approximately 2.2 pounds in 1 kilogram,

$$1.77 \frac{\text{lbs}}{\text{litre}} = 1.77 \frac{\cancel{\text{lbs}}}{\text{litre}} \times \frac{1 \text{ kg}}{2.2 \cancel{\text{lbs}}} = \frac{1.77 \times 1}{2.2} \frac{\text{kg}}{\text{litre}} = 0.8 \frac{\text{kg}}{\text{litre}}.$$

Not much of a difference? Using the factor for pounds per litre (1.77) instead of that for kilograms per litre (0.8) meant that the plane had less than half the fuel it should have had in its tanks. Somewhere near Winnipeg both fuel tanks ran empty and the engines shut down. The only option was to make an emergency landing at the Gimli airbase, on an old runway that had been converted to a racetrack. With no engine power, the plane glided to a rough yet safe landing. A wonky conversion had resulted in a near-death experience for all on board.

A DOLLAR SAVED IS . . . A DOLLAR SAVED!

Understanding conversion factors is also important when I go to the bank. On good days, I sometimes have American cheques to deposit into my Canadian account. The tellers need to convert the US dollar amount into the Canadian amount. This used to be a happier event when the Canadian dollar was low compared to the US dollar, but nowadays, there's only a small difference. Still, I regularly follow the rates, which are conversion factors. (Dollar amounts in different currencies are clearly proportional, as doubling or tripling the amount you have in one currency doubles or triples the amount you have in the other currency.) The rates quoted today, for example, are as follows:

$$\$1 \text{ CAN} = \$0.9494 \text{ US}$$
$$\$1 \text{ US} = \$1.0533 \text{ CAN}$$

I don't need to know both rates because I can derive one from the other. Back in university, I had a professor who told me that he went into mathematics because he had a bad memory. Mathematics is unique as a field of study in that you don't have to memorize much; you can derive what you need to know from a small set of facts. How mathematics is often taught in school flies in the face of this reality. Most people avoid mathematics precisely because they feel there is so much to memorize. But this is mathematics without understanding. Mathematics with meaning places fewer demands on memory than practically any other discipline.

If we know that $1 CAN is equal to $0.9494 US (that is, 94.94 cents US), then

$$\$1 \text{ US} = \$1 \ \cancel{US} \times \frac{\$1 \text{ CAN}}{\$0.9494 \ \cancel{US}} = \$\frac{1 \times 1}{0.9494} \text{ CAN} = \$1.0533 \text{ CAN}.$$

(You may sometimes find a small discrepancy between the two rates, but they should be pretty to close to what you calculate.)

The cheque I plan to deposit is for $1,800 US, so using my conversion factors, I know ahead of time that the amount deposited into my Canadian account should be

$$\$1,800 \text{ US} = \$1,800 \text{ US} \times \frac{\$1.0533 \text{ CAN}}{\$1 \text{ US}} =$$

$$\frac{\$1,800 \times \$1.0533 \text{ CAN}}{1} = \$1,895.94 \text{ CAN.}$$

Actually, I know that when I go to the bank I'll get slightly less, as the bank shaves a few points off the conversion factor for itself; it "buys" the cheque from me at a lower conversion rate, "sells" the cheque at a higher conversion rate, and pockets the extra cash. I might actually be better off holding on to the cheque for a while. Here's hoping that the Canadian dollar tanks . . . Oops, sorry!

TIME IS ON MY SIDE

Conversion factors can be used in a lot of situations where the units are not related, but what you are given is a *rate*. For example, the download rate on a computer is usually listed as so many kilobytes per second. You can view this as a conversion from file size to time. On my computer, the download rate, with a cable modem, is about 350 kilobytes per second, so the conversion factor is 350 KB/1 sec. That's fine, though most files are now measured in megabytes, not kilobytes. But the great thing about conversion factors is that they can be linked to form new ones. There are 1,024 kilobytes in one megabyte, so we get two conversion factors: 1,024 KB/1 MB and 1 MB/1,024 KB. We want to convert our conversion factor of 350 KB/1 sec into one with kilobytes replaced by megabytes. To do this we need the kilobytes to cancel out, so the conversion factor we multiply 350 KB/1 sec by has to have kilobytes in the bottom:

$$\frac{350 \text{ KB}}{1 \text{ sec}} \times \frac{1 \text{ MB}}{1024 \text{ KB}} = \frac{350 \times 1 \text{ MB}}{1 \times 1024 \text{ sec}} = \frac{0.34 \text{ MB}}{1 \text{ sec}}.$$

The file I'm thinking of downloading is 142 MB in size, so I can use the conversion factor (flipping it upside down so the MB cancel) to quickly figure out about how long it will take to download:

$$142 \text{ MB} \times \frac{1 \text{ sec}}{0.34 \text{ MB}} = \frac{142 \times 1}{0.34} \text{ sec} = 417.65 \text{ sec}.$$

That's fine, but I'd rather see the answer in terms of minutes. How can we do that? Conversion factors, naturally!

$$417.65 \text{ sec} \times \frac{1 \text{ min}}{60 \text{ sec}} = 6.96 \text{ min}.$$

I now know that I can go grab a coffee and by the time I get back the download will be finished.

KITCHEN CONVERSIONS

The kitchen is a perfect place to apply conversion skills. Since I've found out that I can follow recipes fairly well, I have taken to trying new things in the kitchen. I view each amount in the recipe as a conversion factor. For example, in my black bean soup I add in four cups of vegetable stock. The recipe serves eight. The conversion factor is 4 cups/8 servings. The dinner party my wife and I threw last week was for seven. So to adjust my recipe, I needed to convert seven servings to the number of cups of vegetable stock needed. Conversion factors did the culinary trick:

$$7 \ \cancel{\text{servings}} \times \frac{4 \ \text{cups}}{8 \ \cancel{\text{servings}}} = 3.5 \ \text{cups.}$$

This may seem obvious to some, but knowing how to use conversion factors can save you from doubling an ingredient when you should have halved it. I have – I mean I know people who have made that mistake. I remember reading that Liberace was once cooking lasagna and instead of grabbing the parmesan shaker, he grabbed a Comet cleanser shaker. Even I cook better than that! Though I always have guests try my food first, just in case.

AND SPEAKING ABOUT HEALTH . . .

Conversion factors crop up in a variety of ways in the health field. After being at home while our sons were young, my wife had to recertify as a nurse. On a number of tests there were questions about correct dosages of medications, and conversion factors were exactly what the doctor ordered.

Drugs are usually prescribed according to the patient's weight (there can be some exceptions, so be careful!). One day my 12-year-old son was at home with me, and his bug bites from camp were driving him crazy. The problem I had was that his weight was off the scale for children's over-the-counter antihistamine. Both of my boys are exceptionally tall, and my 12-year-old is already six feet. (I get *his* hand-me-downs now, and I seem to be better dressed than before.) On the bottle's dosage table, I read the following:

WEIGHT	DOSAGE (EVERY 6 HOURS)
Under 24 lbs or 11 kg	2.5 ml
24-48 lbs or 11-22 kg	5 ml
48-95 lbs or 22-44 kg	10 ml

My son weighs 145 pounds – all muscle and breakfast cereal. The conversion factor I used between millilitres of antihistamine and body weight was 10 ml/95 lbs (I used the upper value for the weight range from the table since I wanted to be safe). I dosed my son with 145 lbs × 10 ml/95 lbs = 15 ml, or three teaspoons. I didn't even need to call my wife.

And while we're talking about health, I have been trying to improve my eating habits. When my siblings and I were growing up, my mother taught us that in planning a meal, you first decide on dessert, the most important part; the rest of the meal is incidental. This hasn't been so good for my cholesterol levels. So I have taken to reading food labels very carefully, watching my intake of various nutrients. Here again, conversion factors are useful, in that every food label lists the nutritional components *per serving*. The can of tuna I'm holding says that one serving equals half a can, drained, or 60 grams, and contains 0.5 g of fat and 170 mg of sodium (another thing to watch out for, if you want to keep your blood pressure down). These are essentially conversion factors.

	AMOUNT	% DAILY VALUE
FAT	0.5	1%
Saturated	0 g	0%
Trans	0 g	
CHOLESTEROL	20 mg	
SODIUM	170 mg	7%
CARBOHYDRATES	0 g	0%
Fibre	0 g	0%
Sugars	0 g	0%
PROTEIN	16 g	

I find that half a can of tuna is not quite enough for me; I need about three-quarters of a can, which is one and a half servings. If you don't see why the latter is true, use conversion factors:

$$\tfrac{3}{4} \text{ can} \times \frac{1 \text{ serv}}{\tfrac{1}{2} \text{ can}} = \frac{\tfrac{3}{4} \times 2}{1} \text{ serv} = 1.5 \text{ serv.}$$

(Back in school you learned that dividing by a fraction is the same as multiplying by the *reciprocal* of the fraction, so that 1 divided by ½ is the same as 1×2/1, which is 2.)

Again using conversion factors, the amount of fat and sodium in my one and a half servings of tuna are

$$1.5 \text{ serv} \times \frac{0.5 \text{ g}}{1 \text{ serv}} = 0.75 \text{ g and}$$

$$1.5 \text{ serv} \times \frac{170 \text{ mg}}{1 \text{ serv}} = 225 \text{ g respectively.}$$

What is my daily allotment of fat and sodium? I could memorize this, but it is there on the nutrition label, too, though I need a bit of math to uncover it. The label, in addition to listing amounts per serving, lists the percentage daily value per serving. The percentage daily value of fat is 1%. That is, 0.5 g of fat, the amount it lists as the fat per serving, is 1% of my total daily fat intake. If I let f be my total daily fat intake, in grams, then 1% of f should be 0.5 g. Recalling that "of" translates into multiplication and that to convert percentages to a decimal we move the decimal place two positions to the left, we get $0.01 \times f = 0.5$ g, so that $f = 0.5/0.01g = 50$ g. So 50 grams of fat is my daily allotment of fat. Actually, this is the daily allotment for women; my manly allotment is a little higher, 75 g. The label, not wanting to be gender specific, uses the lower value.

Similarly, the percentage daily value of sodium, 170 mg, in one serving is 7%, so my daily allotment of sodium is $170/.07$ g = 2,429 g, which is about 2,500 g. Eating the three-quarters of a can of tuna isn't going to bust the bank, so it's worth chowing down on it.

I'VE GOT THE BEAT

I've loved music ever since I was little. I started off on violin when I was really tiny, and after a few years I started taking piano lessons. It's well known that there is a correlation between learning mathematics and learning music. Children who listen to music early tend to do better at mathematics. But this really shouldn't be too surprising. Both music and mathematics involve patterns, as I discuss in Chapter 12. Musical patterns are often appreciated at a subconscious level. Interesting, mathematically based musical sequences find broad appeal, even among those who have no appreciation for mathematics.

I work out at a gym five or six days a week. I am fairly regimented about my workouts, and I listen to some of my favourite music while on the treadmill. I've just read that you get a better workout if the beat of the music matches the tempo of the workout. I especially favour early Beatles songs. A quick calculation tells me why these work so well during my treadmill jog. I do a fast walk at 4.0 miles per hour. First, to convert this to steps per minute, I need to know how many feet are in a mile, 5,280, and then I need to convert feet to steps (or strides). After asking my wife to hold a measuring tape on the floor (we've been together so long she no longer asks why when I ask her to do strange things), I measure my stride at about two and half feet. Thus, using conversion factors, my pace on the treadmill is about

$$\frac{4.0 \text{ mi}}{1 \text{ hr}} \times \frac{1 \text{ hr}}{60 \text{ min}} \times \frac{5,280 \text{ ft}}{1 \text{ mi}} \times \frac{1 \text{ stp}}{2.5 \text{ ft}} = 140.8 \frac{\text{stp}}{\text{min}}.$$

All the glorious cancelling ensures that I have converted in the correct way.

Many of the early Beatles songs, like "I Want to Hold Your Hand" or "A Hard Day's Night" have a tempo of around 140 beats per minute, so they are ideal for my workout. For those songs that are a bit faster or slower, I naturally adjust my stride so that I'm right in time with the music. *Yeah, yeah, yeah!*

CONVERTING YOUR GAS

One of my brothers is a superb musician, incredibly gifted. But like many who are cursed with such talent, he has had to go out and get a "real" job, and he works in accounting. He called me last night with a problem. He has a client who wants to know if he might be missing some receipts for gasoline expenses for the previous tax year. He knows that he travelled about 28,000 kilometres that year. Could I help decide if there were some gasoline bills yet to be found? Even though it didn't appear to be a problem about conversion factors, I could view it as such – converting dollars spent on gas to kilometres travelled.

The client has receipts for $1,500 in gas over the year. Taking a rough average rate of $1.05 per litre over the year, I could convert the dollars spent to litres. Remember, "per" translates as division, and we need to write the conversion factor so that the units cancel properly:

$$1,500 \ \cancel{\text{dollars}} \ \times \ \frac{1 \text{ litre}}{1.05 \ \cancel{\text{dollars}}} = 1,429 \text{ litres.}$$

Now we need to convert litres to kilometres, and I must find out how many kilometres per litre the vehicle gets. I go online but the only information I can find is that this model of car gets about 26 miles per gallon. Not a problem: I can convert between miles and kilometres and gallons and litres. I quickly find on a measurement conversion website that 1 mile = 1.61 kilometres and 1 gallon = 3.79 litres (approximately) so,

$$\frac{26 \ \cancel{\text{mi}}}{1 \ \cancel{\text{gal}}} \ \times \ \frac{1 \ \cancel{\text{gal}}}{3.79 \ \cancel{\text{litres}}} \ \times \ \frac{1.61 \text{ km}}{1 \ \cancel{\text{mi}}} = 11.04 \ \frac{\text{km}}{\text{litres}}.$$

Now for the final conversion: $1,429 \ \cancel{\text{litres}} \times 11.04 \ \frac{\text{km}}{\cancel{\text{litre}}} =$ 15,776 km. My estimation for the number of kilometres travelled over the year is 15,776, based on the receipts that my brother's client had in hand. Given that the client was a salesman who travelled about 28,000 kilometres per year, I told my brother that there are other receipts that his client needs to find.

I find that people often get lost in the calculations and multiply when they should divide, or divide when they should multiply. Have you ever had a brain freeze and been unable to remember how to calculate the distance you travelled from the time it took and the speed you were travelling? Is distance speed divided by time? Speed times time? If you check the units, speed will be measured in say km/h and time in h (hours). If you divide speed by time the units will be

$$\frac{(\text{km/h})}{\text{h}} = \frac{\text{km}}{\text{h}} \times \frac{1}{\text{h}} = \frac{\text{km}}{\text{h}^2}$$

(Remember that to divide a fraction by a number, you invert the bottom and multiply.) These units are definitely not those for distance, so that formula can't be right. You pretty soon see that only speed (km/h) times time (h) gives you the correct units, $(\text{km}/\cancel{\text{h}}) \times \cancel{\text{h}} = \text{km}$, so distance = speed × time is the correct formula. There is a whole branch of mathematics called *dimensional analysis* that deals with finding formulas by keeping track of the units, and it's a very powerful technique – one that you can start using now.

WHEN YOU CAN'T USE CONVERSION FACTORS

It might seem that you can always use conversion factors to convert from one unit to another, but there are exceptions. The fact that there is a conversion factor between one unit and another means that if we start out with *none* of some quantity measured in the first units, we must also have none when we measure in the second units. Zero kilometres is the same as zero miles. Zero kilograms is the same as zero pounds. To be able to use conversion factors, you must have the zeros line up. Isn't this always the case? Nope – temperature is an example, right? Zero degrees Celsius is 32 degrees Fahrenheit, not 0 degrees Fahrenheit. So converting temperature from Fahrenheit to Celsius (or the other way around) can't be done using conversion factors alone.

But that doesn't mean it can't be done. What you need is the relationship between degrees Fahrenheit and Celsius, given by the formula $F = 9/5\ C+32$. Here F is the temperature measured in degrees Fahrenheit and C is the same temperature measured in degrees Celsius. I usually remember it in words: to convert a temperature in Celsius to Fahrenheit, I divide by 5, multiply by 9, then add 32. The order of operations is important; multiplication and division must come before addition and subtraction unless brackets force you to do otherwise.

So, to convert 20°C, we divide 20 by 5 to get 4, multiply by 9 to get 36, and add 32 to get 68°F. When I do this in my head, I often round off the division by 5 if it doesn't turn out to be a whole number.

To convert from Fahrenheit to Celsius, I don't bother to memorize a new formula. Too much overhead! Instead, I simply write down the equation and use a bit of algebra to get C by itself on one side:

$$F = \frac{9}{5}C+32$$

$$F-32 = \frac{9}{5}C$$

$$5(F-32) = 9C$$

$$\frac{5(F-32)}{9} = C$$

All of these steps use the basic algebra you were taught in elementary and junior high school, but you may not have used it for a very long time. The bottom line gives a formula for converting temperatures in Fahrenheit to Celsius: subtract 32, then multiply by 5, and divide by 9.

The black bean soup for the dinner party turned out great. All the appropriate recipe conversions were accounted for, and the smoke detector didn't go off. The soup actually tasted not half bad, so I just had to remember to serve that half. With a successful dinner under my belt, I eyed a dessert recipe for a flambé. I had never tried making one before, but the barbecue lighter seemed easy enough to use. What's the worst that could happen?

3

ONE GRAPH IS WORTH
A THOUSAND WORDS

The house is pretty quiet now, after the boys have headed off to school. I take a few moments and settle down with a cup of decaf coffee to read the paper. Long ago I gave up caffeinated coffee, the drug of choice among mathematicians. One of the greatest mathematical minds of the past century, Paul Erdös, said that a mathematician was a machine that converted coffee into theorems (mathematical results). Baked goods used to work for me, though I have had to go cold turkey on those too, for cholesterol reasons. If I ever have to take a drug test, the only brain-enhancing substance they will find in my system will be omega-3s.

Looking through the newspaper, a variety of visuals catch my eye, from photographs of politicians trying to explain away their latest gaffes to an array of graphs in the business section. I've always liked graphics. In high school I loved art class, and without a doubt it was the visual nature of my area of mathematics – graph theory – that drew me in like Ulysses to the Sirens. Numbers are nice, but there's nothing like a good plot.

Almost everyone thinks of mathematics as being exclusively about numbers. I can't tell you how many times people, upon hearing I am a mathematician, exclaim, "Oh, you must love numbers!" I doubt many people think that carpenters love nails or architects love lines. But such is the false image that the outside world has of mathematics.

Most mathematicians visualize the mathematical objects they work with. Mathematics gets a bad rap as some sort of symbol-pushing – a Greek symbol here, a Greek symbol there. I know many students who have claimed that mathematics is "all Greek to me," and not without just cause. But this view of mathematics – as a special kind of language, to be manipulated in some mysterious way, only to have the correct answer pop out at the end – is widespread. Mathematics may be taught this way by teachers who lack confidence in mathematics, who fail to visualize mathematics, and even by those who know better but who are not willing to put into their teaching the effort they put into their research. But there are many of us who see ourselves as mathematical ambassadors, who wholeheartedly want to share our love for and intuition in mathematics. So I am going to take you on a tour of our underdeveloped mathematical country, and point out the lay of the land, as I see it.

DATA, DATA EVERYWHERE

When I open the newspaper, the one thing that catches my eye is the sheer amount of data. The stock market pages are a massive set of data with a dizzying list of volumes, closes, +/-, highs, lows, and so on, without any indication of what the numbers mean. It is difficult, even as an analyst, to make sense of a lot of data. One way to condense the data is to use statistics, which I talk about in the next chapter. Another way is to capture the essence of the data in a picture. One picture is indeed worth a thousand or so data points. The trick is finding the right picture to go along with the data, one that highlights what we are interested in, rather than diverting our attention.

In investing, what probably is most important is not the current price of an individual stock, but its trend over time. Has the price of the stock been climbing or dropping? Do I think the trend will continue or change? The information on the stock page gives only a snapshot of what is happening now. True, the +/- column indicates how a stock's price has changed since yesterday, but this I consider mainly noise – a small change that may be due to randomness. Focusing on daily fluctuations may hide whatever real trend is going on. After all, a stock's price from day to day can be affected by any number of things, and small changes one way or the other may mean nothing.

Here are some typical share prices over 10 days: $11.50, $11.38, $11.59, $11.47, $12.14, $12.45, $12.79, $13.52, $13.87, $14.24. If I want to visualize the data, I would create a plot with a point for each day, with the height of each point being the corresponding price of a share of the stock. I wouldn't bother listing the actual date of each price, just the number of days since the initial $11.50 price, which I would put at "day 0." Typically, mathematicians start numbering at 0, though most normal people start at 1. (I like it that in Halifax the main highway starts with exit 0 rather than exit 1.)

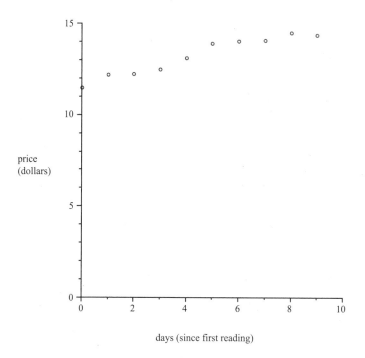

days (since first reading)

I might reduce the size of the points and join adjacent points by straight lines to form a *line graph*. Many plots of stock prices are done this way, though it does give the false impression that in between the close of each day the value of the stock changes gradually (and constantly) from one value to another.

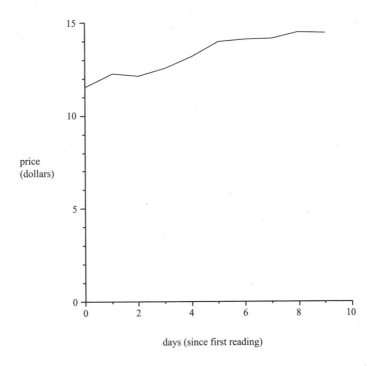

days (since first reading)

The plot makes apparent some things that may not be obvious from the data. For while a stock's price may bounce up and down a little throughout the 10 days, the graph shows that the trend is clearly upwards. If I believe that trend will continue, it would be good to buy now in order to sell later. If the trend were downwards, I could sell the stock "short," an investing strategy where you sell something you don't actually own (not recommended with garden tools borrowed from neighbours). Of course, at some point in the future you have to agree to purchase enough of the stock to cover what you have sold, so you are banking that the price will have dropped further by then.

But the graph of the points shows more than whether the trend is up or down. It reveals how much a stock's price changes daily. I can see that the points seem to form roughly a straight line, angling upwards. Seeing straight lines is just something we humans are inherently good at. We are much, much better at seeing a straight line than at judging whether a set of numbers has a linear relationship. In fact, I think I can draw a pretty good line through the points just by connecting the first and last point.

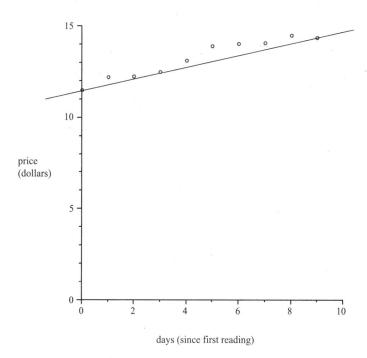

days (since first reading)

I could probably find a better line through the points by joining other pairs of points. Is there a best line I could draw through the points? See, now you're thinking like a mathematician! Don't be scared – that's a good thing. Indeed there is a best line. The process of finding the best-fitting line is called *linear regression*, and statistical and spreadsheet programs perform these calculations.

The slope of the line in the diagram tells me how much the price of the stock changes daily. I remember how to calculate the slope from elementary school (it's one of a handful of useful facts we are taught in school whose importance in everyday life generally goes unappreciated). It's the "rise" over the "run," where the "rise" is how much you go up and the "run" is how much you go across. From the first to last point we go up from 11.50 to 14.24, so the rise is 14.24–11.50=2.74. The run is how far we go horizontally, which is 9–0=9 (from day 0 to day 9). So the slope of the line is 2.74/9=0.30. This means that over a large number of days, the stock seems to rise by about $0.30 per day. Over 100 days, I would expect the stock to rise about 100×$0.30=$30 from its initial opening of $11.50, provided that the trend continues that long. All of this predictive power, merely from plotting the data!

THE EYES HAVE IT

Today we have so much data at our fingertips – not hundreds or thousands of numbers, but millions and billions, and this makes the ability to visualize especially crucial. But there is an art to learning how to read graphics appropriately. How a banker, an engineer, or a political party visually presents data depends on what point they are trying to make. Pie charts, for instance, emphasize the relationship between the parts and the whole. The size of each piece of a pie should be proportional (in the sense we discussed in Chapter 2) to the percentage that piece is of the whole. Poll results for a multiparty political system such as we have in Canada are a perfect place to use pie charts. In a recent online poll, the following numbers were given: support for the Conservatives was at 34%, Liberals 31%, NDP 16%, Bloc

Québécois 10%, and Green Party 9% (note that the sum of all the percentages is 100% – I would guess that the poll recorded only voters who felt strongly enough to have party preference). The whole pie (a circle) consists of 360 degrees. So in a pie chart, the 34% support for Conservatives corresponds to a piece of the pie with 34% of 360 degrees. Remember that "of" means multiplication and that 34% is 0.34, as a decimal. So the angle for the Conservative piece of the pie should be 0.34×360=122.4 degrees. You calculate the other pieces of the pie accordingly to get a picture like the following:

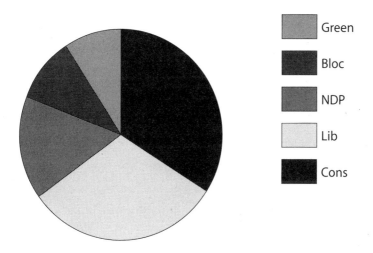

- Green
- Bloc
- NDP
- Lib
- Cons

In a glance you can see that the support for Conservatives and Liberals is about the same, and twice as much as that for the NDP. Neither the Conservatives nor the Liberals can garner greater than 50% support, even if they form a coalition with one of the other parties (besides each other). The numbers tell the same story, but mixing the visual with the mathematical makes the numbers come alive.

Pie charts are good for relative comparisons – how large each piece is relative to the others – but what gets hidden are the actual numbers themselves. If what you are interested in are the numbers themselves, then histograms are the way to go. Histograms are diagrams consisting of vertical bars, where the height of each bar represents the relative size of the corresponding number.

Tax rates in Canada seem inordinately high compared with those of many other countries, especially our neighbours to the south. The Canadian government would like each of us to forward all of our earned income as tax, but settles for slightly less.

Here in Canada we have a sales tax known as the GST, the goods and services tax, though there is nothing good about it. I did a little investigation and found out, much to my chagrin, that many other countries have even higher sales taxes, known as VAT, value added tax, though I have yet to see a tax that actually adds value to what I am thinking of buying! The United States doesn't have a VAT, and therefore its bar in the histogram has no height, and hence is invisible. That's one of the drawbacks to visual models like pie charts and histograms – you can't see what isn't there.

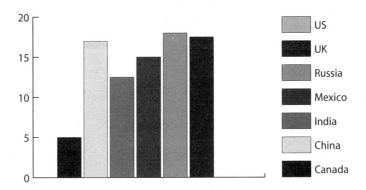

VALUED ADDED TAXES AS PERCENTAGE OF SELLING PRICE

LIAR, LIAR, GRAPHS ON FIRE

How can graphics lie? It's easy if you want them to. I'm going to show you a few tricks, not so that you can use them, but so that you can be aware when someone is trying to pull the graphical wool over your eyes.

Here's a simple example. A company's sales for the past two years are shown below in a pretty graphic. The graphic is clearly meant to mislead; the one on the left has six money bags, while the one on the right has five, so the fact being reported is that sales are down (by a sixth, or about 17%). On the other hand, the vertical stacking of the five bags on the right might lead the viewer to think at first glance that sales have gone up. The natural corre-lation between a dimension of height and increasing value has been manipulated.

last year   this year

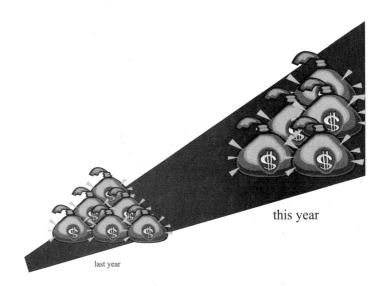

this year

last year

Three-dimensional graphics have become the rage but looks can be deceiving. Here the path to riches has perspective attached. Captivating? Perhaps. Confusing? Without a doubt! Again, we are tricked by our eyes into believing that there is growth from last year to this year, when in fact the opposite is true.

There are even more subtle ways in which graphs can manipulate our judgment. Suppose that sales at a company for the past three years were $315,000, $316,000, and $318,000. Basically, there has been no significant change over the three years, which, if you factor in inflation, might be a big problem. The following line plot shows the sales over the three years.

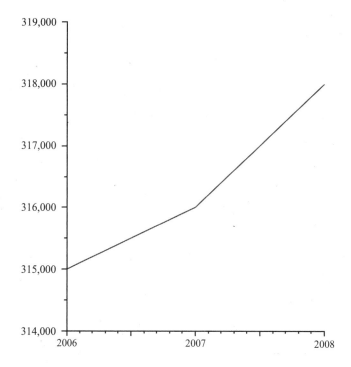

This plot is completely misleading as it appears to show a significant growth in sales. The problem lies in the fact that the bottom of the graph starts not at 0, where we would expect it to, but at $314,000. The changes from one point to another seem much larger than they really are, since the changes should be shown not in comparison to each other, but to the total sales. A more realistic plot would look like the following:

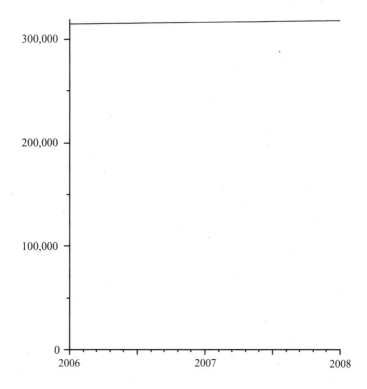

So instead of dropping off to sleep during PowerPoint presentations, you might want to stay awake and look at the graphics with a careful eye; there may be a lot at stake.

THE PROOF IS IN THE PICTURE

I use pictures all the time at work, whether I am teaching or doing research. This is partly my artistic side wanting to come out, but it also stems from my desire to cultivate my understanding of the mathematics, to make the mathematics more real. When I teach calculus, I fill the blackboards (and nowadays whiteboards, overheads, and LCD projectors) with a variety of different graphics. Students always ask me whether they *need* to draw pictures on assignments and tests. My answer is always the same: "No, but then again, you don't *need* to get the correct answers either." Without pictures, mathematics often becomes a lot of meaningless symbol manipulation, where one answer is as good as any other.

Calculus, for example, at the most basic level, has to do with two things – the rate at which quantities change and areas. Here is a favourite problem of mine:

The following shows a plot of the function $f(x)=2x^2e^{-x^2}$ over the interval $[0,4]$. Show that the shaded region has an area of at most 3.2 square units.

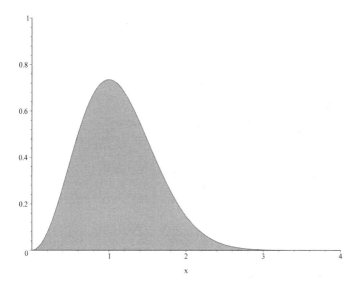

There is a whole branch of calculus called *integration* which calculates areas based on formulas, but in this case I've carefully chosen the formula so that such a procedure is bound to fail. I've tossed in the proverbial red herring.

On the other hand, just look at the picture. The shaded area is certainly less than the area of the rectangle (whose width is from 0 to 4 but whose height is 0.8), since the shaded area sits completely within this rectangle.

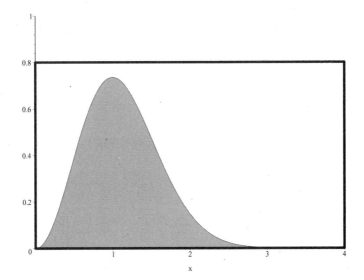

The area of the rectangle is its width times its height, which is 4×0.8=3.2, so we conclude that the shaded area is at most this value.

This example highlights what I want to illustrate – that mathematics is more than just calculations. It is insight, a creative way of thinking. We could all take this lesson to heart in our everyday lives.

The calculus example shows that arguments can be easy if only we look at them in the right way. And there is nothing that mystifies people more about mathematics than proofs. There is a certain fear factor associated with the word *proof.* But we are all accustomed to requiring proof, whether it's guilt in a court of law or to settle an argument with a friend. Mathematics simply demands validation before the truth of a statement is accepted.

Let me give an example. Suppose you want to add up all numbers from 1 to 100. $1+2+3+4+\ldots+97+98+99+100=?$ There is a famous story about a teacher who, being annoyed with his class of eight-year-old boys, sat them all down to add up the first 100 whole numbers. He expected this problem to keep them busy for a long time, but little Carl Gauss jumped up almost immediately with the correct answer, 5,050. How did he do it? He noticed that in the sum $1+2+3+4+\ldots+97+98+99+100$, if you add the first and last numbers, 1 and 100, you get 101. You get the same result if you add the second and second-last numbers, 2 and 99, and so on, up to the fiftieth and fiftieth-last numbers, 50 and 51. So the sum breaks down into 50 pairs of numbers, each adding to 101. So the final answer is 50×101, which is 5,050. This was only the tip of the iceberg for Gauss; he went on to become one of the greatest mathematicians of all time.

You can use mathematics to show that if you add up the first n numbers, $1+2+3+\ldots+n$, then what you get is always $\frac{n(n+1)}{2}$. So $1+2+3+4+\ldots+97+98+99+100=\frac{100\times101}{2}$, which is 5,050. Mathematicians like to generalize, to prove statements and formulas in the broadest setting. Carl Gauss's argument can be used to prove this, though some care has to be taken when the number of numbers you are adding up is odd. But there is a beautiful little picture proof that tells the whole story:

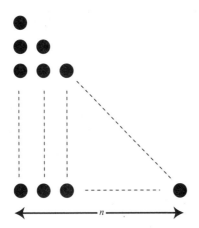

I've placed a set of balls to represent each number – one ball in the first row, two in the second, and so on, with n balls in the last. (We mathematicians are used to putting in dotted lines for missing objects, with the understanding that they stand for some fixed number of things.)

Anyway, the sum $1+2+3+ \ldots +n$ is simply the number of balls in the picture. Now we pull a little trick and flip the picture upside down, and move the two pictures together as shown below:

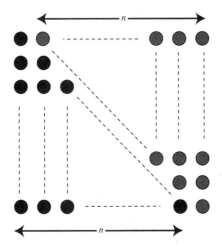

Each of the n rows has exactly $n+1$ balls in it. So if we add all of the balls in all of the rows we must get $n\times(n+1)$. On the other hand, this total number of balls is twice the number of black balls (as there are just as many grey balls as black ones), so the number of black balls is half of the total number of balls, that is $\dfrac{n\times(n+1)}{2}$. This is exactly the sum $1+2+3+\ldots+n$.

Ta da! – a proof that is essentially in the pictures. Is it a *real* mathematical proof? Absolutely – it is completely convincing, and the bonus is that we can literally see how it works.

SOME MORE PICTURE-PERFECT PROOFS

Here is another favourite proof of mine. Suppose you take a checkerboard and a set of dominos. If you place dominos either horizontally or vertically, to cover two adjacent squares of the checkerboard (and no two dominos can overlap, naturally) you can cover the entire checkerboard with dominos (just use four dominos in each row). But now suppose you eliminate the end squares in two opposite corners. Can you still cover the squares with the dominos?

You could try taking an actual set of dominos and a checkerboard and covering all the squares except the two opposite corners, but I warn you ahead of time that you are bound to fail. Why? By drawing the picture, I see a couple of things. First, the two squares I throw out, being at opposite corners of an 8×8 checkerboard, are the same colour – just look! Secondly, a domino covers a white and a black square, no matter whether it is placed horizontally or vertically. But this implies that the number of white squares covered by the dominos must be equal to the number of black squares covered by the dominos. There are 32 white squares and only 30 black squares to cover, so it can't be done, no matter how hard you try!

YOU KNOW THE GAME

Here is a game you probably tried when you were young. The object is to trace the figure below, starting anywhere you like, without taking your pen off the page, and without retracing any part of a line. I remember spending a good part of many childhood days trying such puzzles.

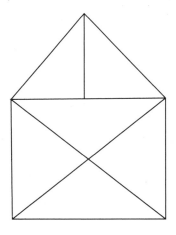

The figure shown here is impossible to trace. Why? Place a small dot at each crossing or corner.

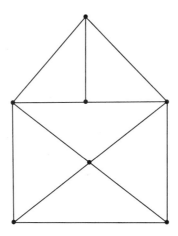

If the figure is traceable, you should be able to convince yourself that you can do so by moving from point to point.

The figure shown below represents a general tracing.

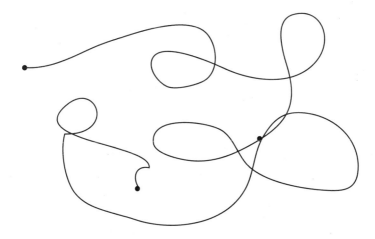

For every point but the start and finish, there must be an even number of lines coming out of it. The start and finish, if they are different, will have an odd number of lines coming out of them, and if the start is the same as the finish, all points will have an even number of lines coming out of them. So if you can trace the figure, either none or exactly two points will have an odd number of lines coming out of them.

This is why the figure we started with couldn't be traced; there are four points with an odd number of lines coming out of them.

Such figures, with dots joined by lines, are called *graphs* (or *networks*) and they form my particular area of research. (I am known as a *graph theorist*, which sounds a lot better than "the man who draws dots and lines.") The particular problem of tracing a graph (without retracing your steps or lifting your pen from the page) is known as finding an *Eulerian trail*. Beyond recreational uses, it has applications in scheduling things like garbage pickup routes. One of the great things about mathematics is its recycling property – if you are lucky and you look hard enough, you may be able to reuse some mathematics, yours or someone else's.

Can all figures that have either two or no points with an odd number of lines coming out of them be traced? The answer is . . . yes, provided that the figures are, of course, in one piece. If there are two points with an odd number of lines coming out of them, choose one of these to start with; otherwise, choose any point to start. Then imagine that you erase lines as soon as you walk over them. The rule is that you continue your walk over a remaining line, provided, if possible, that erasing the line doesn't break up the remaining figure into more parts. This procedure, known as *Fleury's algorithm*, is guaranteed to work, always. Give it a try with the following figure.

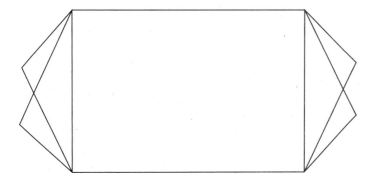

I READ THE NEWS TODAY, OH BOY (AND GIRL)

I have just read in the newspaper the latest statistics said to detail how many sexual partners heterosexual men and women have in their lifetimes. The rough estimates suggest that men have, on average, seven partners, while women have four. But I, like some other mathematicians, turn up my nose at such data. Is this reaction because most mathematicians would likely rank well below the average? (We're lucky if we can get *any*.) But no, the problem is deeper than that.

Here is what I see when the figures are presented to me as fact. I see the heterosexual men and women as points in a graph, men on one side, women on the other. Then in my mind I draw a line between two points, if a man and a woman have had a sexual relationship with each other. The picture is something like this:

men

women

(The wavy line is a representation of lines that are there but not specified.)

The number of lines coming out of each point (which is called the *degree* of that point) is the number of partners for that person. If we add up the number of lines (i.e., sexual relationships), then divide by *m*, the total number of men, we get the average number of partners for men, which according to the newspaper article is about seven. So the number of sexual relationships is the average number of sexual partners for men times the number of men, that is, $7m$. Likewise, the number of sexual relationships for women is the average number of sexual partners (given in the article as 4) times the number of women (which we'll call *w*), that is, $4w$. So $7m=4w$, which implies that $m=4/7w=0.57w$. This means that the number of heterosexual men in the world would have to be only about 57% of the number of heterosexual women, which is absurd.

So what gives? Certainly, there might be a problem in terms of those who are surveyed. Some people who have been in relationships will have died, but it's likely to be roughly the same for both genders, so death would have little impact on the numbers. The only reasonable inference is that the math doesn't lie – people do! Men are likely to increase their numbers, while women likely under-report, on average. As for the mathematicians surveyed, there may well be a difference between theory and practice!

I have finished reading the newspaper and I have filled a couple of pages with some doodles of math research. I rip off the important pages, and stuff them in my back pocket to look at again later. Boy, I'm glad I wrote over only the sports section – my wife rarely bothers to read those pages. The last time I scribbled over the front section, it wasn't a pretty picture, I tell you.

4 JUST THE STATS, MA'AM

I'm off to work at the university. It's a relatively short drive in, just 25 minutes, 30 minutes in morning rush hour. While I idle in a traffic jam, I let my mind drift back to the research problem I've been working on. I know that occasionally I gesture while I reason things out in my head, which must look very odd to other drivers. I try to be careful about which fingers I use.

I like listening to the radio while I'm driving. Discussion of whether the federal minority government is going to fall fills the news. It seems like every journalist and reporter wants to weigh in. Elections are always full of statistics that are thrown about more than a wrestler on WWF. Yet many people don't have a feel for them. What does the "mean" mean? Should you be anywhere near a standard deviation without protection?

And polls are everywhere. They have an air of trustworthiness about them, but I know that sometimes polls are intended to shape public opinion rather than reflect it. One thing I know is that I'm sick of hearing about politics. Nineteen times out of twenty.

You just can't make sense of the news without some knowledge of statistics. And that's 100 per cent certain.

While some things in our lives are under our control, an equally large number of things are not. Chance plays a role in everything. Whether we will arrive safely at work, whether it will rain later today, whether gas prices will go up at the pumps (again!) – these are just a few events that seem to have an element of randomness, and try as we might, we can't accurately foresee what is going to happen.

My 12-year-old son has started using "random" as an adjective. He and his friends use it as a synonym for *non sequitur*. Replying "antelope" to the question "How are you?" qualifies as "random" in their minds, even if they use the same answer over and over again. But I've tried to convince him that repetition (i.e., predictability) makes this response anything but "random" ("unusual" or "weird" would be a better descriptor).

Randomness is a process by which the outcome can't be predicted with certainty. What probability and statistics do is give us a way to deal with uncertainty in order to make realistic predictions of what is likely to occur, and how likely it is to occur. Nothing in life is certain; there is risk in almost every endeavour. But to live life to its fullest, we need to evaluate the risks and the benefits and make our best choices.

YOU BET!

Some probabilities are easy to determine. Tossing a fair coin is a 50–50 proposition: half the time it comes up heads and half the time it comes up tails. Many games of chance rely on the fact that each possible outcome is equally likely. In poker, each hand of five cards is equally likely to be dealt. In roulette, the ball is equally likely to drop into any one of the 38 pockets. When all outcomes are equally likely, the probability of any one outcome occurring is simply 1 divided by the number of possible outcomes.

The mathematics of such games is well understood. In all casino games the probability of winning is always less than the probability of losing, that is, of the house winning. The *odds* of winning are the probability of winning compared to losing, written with a colon instead of a division sign. The odds of losing are the reverse. Mathematically speaking, if my odds of winning are $m:n$ then my chances of winning are $m/(m+n)$ while my probability of losing is $n/(m+n)$ (and so the odds of losing are $n:m$). So odds of 2:3 indicate that my chances of winning are only $\frac{2}{2+3}$ =2/5, or 40%, while my chances of losing are 3/5, or 60%.

Now the odds offered for winning are not the same as the odds for losing, though they would be in a fair game. Craps is a popular game in which one person, the "shooter," rolls a pair of dice repeatedly. There are a variety of different outcomes to bet on. Let's examine one of these: For rolling "craps," that is, a 2, 3, or 12, on a single roll of the dice (but not the first opening roll of the dice), there is one way to roll 2 (namely ones on each of the two dice) and one way to roll a 12 (two sixes). There are two ways to roll a 3 (a 2 on the first die, a 1 on the second, or vice versa). So in total there are four ways to roll craps. Since there are 36 different, equally likely outcomes for rolling two dice (six choices for the first die and six for the second), the probability of rolling craps on a single roll of two dice is 4/36, which is 1/9, or about 11%. The probability of *not* rolling craps on a single roll of two dice is therefore 1–1/9=8/9, or about 89%. The ratio of not throwing craps to rolling craps is $\frac{8/9}{1/9} = \frac{8}{9} \times \frac{9}{1} = \frac{8}{1} = 8$. (Remember that to divide by a fraction you *invert and multiply*.)

Thus the odds of not throwing a 2, 3, or 12 in a single roll are 8:1. In a fair game your payment if craps came up would be 8:1 as well, that is, for every dollar wagered, you would win $8. But what you will find typically in a casino is that, if you bet on craps coming up in a single roll of the dice, your winnings are 7:1, meaning that for every dollar you wagered you could win $7. Over the long term, repeatedly making the same bet, what you can expect to win is $7, 1/9th or about 11% of the time (when a 2, 3, or 12 does come up) and you can expect to lose $1, 8/9ths or about 89% of the time. Letting amounts we win be positive, and amounts we lose be negative, on average, I'd earn $7 \times \frac{1}{9} + (-1) \times \frac{8}{9} = \frac{7}{9} - \frac{8}{9} = -\frac{1}{9}$, which is about –0.11. (In mathematics, this calculation – adding up the winnings and losses multiplied by their respective probabilities – is called *expectation*.) In the long haul, I can expect to lose about 11 cents on each dollar bet. On any one $1 bet, I might lose the dollar or win $7, but over, say, 100 such $1 bets, I would expect to lose about $11.

It is this spread that keeps casinos in the black. On any individual bet, the casino might win or lose. The probability of this happening is not usually crucial to the casino; what is vital is the payouts that are paid for winning. Casinos make sure that on every type of bet, there is enough of a spread (that is, enough of a difference) between the odds of losing and the payout odds so that, in the end, they make money. That's all there is to it.

Of course, the lower the payout odds are, compared with the odds of losing, the more money a casino would expect to make. So why don't casinos give, say, 5:1 payouts for rolling craps? The key is that casinos want the odds to be in their favour, but only slightly, so that individuals will lose their money more slowly, and feel they have a chance at winning. If the payout odds are 5:1 instead of 7:1 for rolling craps, on average, I'd lose about 33 cents per each dollar bet, rather than 11 cents per bet. I'd run through my money faster, and get less enjoyment from gambling. And the gambling establishments want me to keep coming back.

Is there a way to beat the house? In reality, no. The only way not to lose is not to play!

Having said all this, if you're still hell-bent on betting, you might want to look over the losing odds and payout odds for various bets for different games (there are many sites online that list this sort of thing) and choose the bet that has the smallest spread. Even though you will likely lose money over time, you'll be able to enjoy the experience for longer.

Still, is there a chance you could win? In the short term, yes. Betting on rolling craps in a roll of two dice, you could win three times in a row. The chances of rolling craps on different rolls of two dice is *independent* of the other rolls (the outcome of one roll of a pair of dice in no way affects any other such outcome). In such cases, to get the probability of all of the outcomes happening, you multiply the corresponding probabilities. We've already seen that the probability of getting craps on a roll of two dice is 1/9, so the chances of rolling three craps in a row is $1/9 \times 1/9 \times 1/9 = 1/729$, or about 0.14% of the time. That is pretty darn unlikely, but if you were to come into the casino every night for two years (which is more than 729 days) and roll a pair of dice three times, you could expect on one of those nights to roll craps three times in a row. Suppose you bet $100 on the first roll. You would earn $700 (since the payout odds are 7:1). If you let it all ride on each of the subsequent rolls, you would make $5,600 (betting both the $100 and the $700 winnings, for a total of $800 at 7:1 payout), and finally $44,800 on your bet of $6,400 (the $100 original bet plus the winnings of $700 and $5,600). You would take home a grand total of $51,200, all on an initial bet of $100. Not bad for a night's work!

The fact that rare events eventually occur close to the expected number of times, provided you repeat the experiment many, many times, is what is known as the *Law of Large Numbers*. But while you're likely to have such a night if you gamble every day for two years, what is hidden are the many, many days that you would have lost money. Most days you would have lost at least $100, and your big winnings on that glorious day would be overshadowed by your losses on many other days.

I recently had lunch with a friend who mentioned to me that he had a friend who routinely gambled at the track. This person had one incredible year of winnings, so good that he still talks about it. How did that jibe with the fact that most people lose at the track? I pointed out that we humans are particularly good at remembering unusual events, good or bad. They have extra significance for us. The more ordinary events, especially those that may be a little on the down side, tend to be forgotten over time. Over many years of betting, there is bound to be a year of exceptionally good luck, just as there is bound to be a year of exceptionally bad luck. Probability theory states as much. So it isn't unusual that my friend's friend had a good year. It's just fortuitous that he has forgotten about all the mediocre and poor years.

That strings of good or bad luck are bound to happen is an important fact to keep in mind. We all attach importance to such things, when really it's just dumb luck. Do you know that if you toss a coin 100 times, you are bound to get runs of both heads and tails? You can expect on average to have a run of about six heads, and a run of about six tails. Hold on a second. There. I just tossed a coin 100 times. The tosses started off

T, T, H, T, H, H, H, H, H, T, T, T, T, T, T, H, T, T, ...

Already I see a string of five heads followed by a string of six tails. Is this evidence of something unusual about the coin? Nope. Runs such as these are very likely to occur. It would be unnatural if they didn't. We tend to think of randomness as being evenly distributed, not missing any spots, not covering any spots too much. But true randomness has clumps and sparse spots. It is our interpretation of patterns that is the problem. Sometimes a run is just a run.

I SAY WHAT I MEAN

Games of chance, such as craps, roulette, and poker, are fairly easy to understand mathematically; provided the dice and ball are rolled fairly and the deck is well shuffled, all of the outcomes are equally likely to occur. But there are many events in everyday life that seem on the surface to be random, and yet not all of the possible outcomes are equally likely. Height, weight, IQ, and so on vary amongst individuals, with so many factors entering into play that if you pick a person at random, there would be no way of knowing ahead of time what that person's height, weight, or IQ would be – it could be any number within a range, with some numbers more likely than others.

I don't know about you, but to get a feel for a group of numbers, I need some way of condensing them, to work them down to some values that give me a sense of the whole set. Part of what statistics does is help us to make sense of the uncertainty by summarizing the possibilities with only a few numbers. In one of my classes (with an enrollment of 31), I have to give some summary of the midterm marks to my students. For privacy reasons I can't post the marks. So what is a reasonable value to give as the "typical" mark?

There are a few choices. The most usual one is called the *mean*, which is the average; you add up all of the values, and divide by the number of values you added. So if the test scores were

58, 74, 55, 62, 59, 77, 91, 80, 45, 68, 62, 70, 64, 98, 42, 55, 68, 58, 54, 58, 66, 60, 78, 88, 83, 46, 82, 70, 82, 58, 52

then I would add them all up, to get 2,063, and divide by 31, the number of scores, for a mean of about 66.5. This is the one I post as the "typical score."

People often forget that if there are values above the mean, there have to be values below the mean. Not everyone can be above-average, no matter how utopian it sounds. It just can't be.

TYPICALLY ATYPICAL

But is the mean always an appropriate measure of the typical? If Bill Gates came to live in a small village in Africa, the mean of the yearly incomes would probably be reasonably high. But stating this value as the "typical" income in the village would be completely misleading (though exactly this type of thing is done when the desire is to mislead!).

Things can get even weirder if you group the data. Suppose the employees at your firm are grouped into high-income and low-income earners. Moving the top salary from the low-income group into the high-income group will, paradoxically, decrease the average salary in *both* groups, since you have added a salary that is lower than the salaries in the top group (thereby lowering its average) while removing a top salary from the low-income group, thereby lowering its average as well. Sometimes averages are meaningless, sometimes they are misleading, and sometimes they can be bent to the statistician's will.

If all this is a bit abstract, imagine that you and four friends have spent the evening playing poker. Your friends each lost $1,000, but you came away the big winner, winning $4,000. The average winnings in dollars is $((-1,000)+(-1,000)+(-1,000)+(-1,000)+4,000)/5=0$. (Remember that losing is winning a *negative* amount.) But you'd be hard pressed to say that the typical winnings were zero – not only did nobody come away winning 0, but far from it; they either lost big or won big.

Therefore sometimes the mean is not an appropriate number to promote as the typical value. What often makes sense are two other values, called the *median* and the *mode*. The median is the middle value. To find the median you arrange all of the numbers in order, and choose the middle value (or if there are two middle values, you average them). For the test scores, I would order them as

42, 45, 46, 52, 54, 55, 55, 58, 58, 58, 58, 59, 60, 62, 62, 64, 66, 68, 68, 70, 70, 74, 77, 78, 80, 82, 82, 83, 88, 91, 98.

There are 31 numbers, so the middle value is the sixteenth number. We count from the bottom of the list and find the sixteenth number, which is 64. This is the median of the test scores.

The *mode* is the number (or numbers) that occurs most often. In the test scores, the mode is 58, as it occurs four times, while none of the other numbers appear as often. The median and mode of the poker game winnings are both −$1,000, which I think is a more "typical" amount of winnings in the game.

Which among the mean, median, and mode is put forward is completely up to those who are summarizing the data. I sometimes give all three, though usually I only give the mean.

As if the mean, median, and mode weren't enough, there are also percentiles to deal with. Everyone's probably had some ranking given as a percentile. My sons are exceptionally tall; my younger, when he was a toddler, was in the ninetieth percentile of height and seventy-fifth percentile of weight. Percentiles follow the idea of the median (which is the fiftieth percentile). If you are in the ninetieth percentile, it means that at least 90% of the population (those with whom you are being compared) rank below you.

Percentiles are great if you want to know where you stand in comparison to others. Even if a test is particularly hard or particularly easy, the percentile will tell you how you performed compared with everyone else.

IT'S OKAY TO DEVIATE

Often even the mean, median, and mode are not enough to describe the data. In addition to knowing what the "typical" value is, I often like to know how much variation there is. Recently, our financial advisor came over. In the initial meeting, she asked how much money we had to invest, and after we all stopped laughing, she asked us to complete a survey about our view of money and risk. One of the questions cut quickly to the issue of variation. If we were assured of getting a certain *average* return on our money, would we be all right if there was a considerable amount of variation in the value of our portfolio? That is, would we tolerate large dips (and rises) in the value of our investments along the way, provided the average return was good?

What the advisor was getting at was our view not on the *mean* or average of our investment earnings over time (we would naturally want that to be high), but on what is called the *standard deviation*, which measures how much the data is spread out about the mean (the word *deviate* is to differ, here from the mean). If the numbers are all around the mean, this should be small, but if there are numbers quite far from the mean, it should be large. I won't go into details of how this is calculated, but rest assured, it is exactly what standard deviation measures.

The standard deviation for the test scores, whose mean was 66.5, is about 13.8. The standard deviation of the poker winnings is $2,000, which is quite large, indicative of how much the winnings varied.

My wife and I found out we had slightly different toler-
ances for standard deviations – I could tolerate more, provided
the average return was better while my wife could tolerate less, so
we opted for the portfolio with less variation.

BUT I REGRESS . . .

It is a fact, called *regression to the mean*, that if over time the mean
stays the same, and there is about the same amount of variation,
then extreme values tend to become less extreme. Let me give you
an example. IQ scores, the common measure of intelligence, are
rather static from generation to generation. On average people
don't get any smarter or dumber. So what about children of brilliant people? The children of geniuses tend to be smart, but not
as smart (or smarter than) their parents. You've heard about
Einstein, but what about his children? Moses' children were obviously not in the same class as their dad (rather than passing to
them, leadership went to Joshua). J.S. Bach's kids were good
musicians, but not quite as gifted. If the average intelligence stays
the same over time, it is necessary that the extreme values be less
extreme the next time around. And it works at the other end of
the spectrum, too. I have some acquaintances for whom I've tried
explaining regression to the mean, in order to give them hope
that their children will likely be brighter than they are. But it's all
over their heads.

This discussion brings to mind *Stein's paradox* in statistics,
which states that if you observe the average values of something
over the short run, say points per game for a number of hockey
players, then the best estimates of the long-run averages of each
player are *not* the short-term averages, but rather the short-term

averages modified towards the average of all of the players. It sounds crazy – why should the prediction of an individual's average depend on other people's average? But again, it's a law of statistics. The players who are on a scoring streak tend to score more often than average over the season, and similarly, those in an early scoring drought will likely be below average on the season, but above their present average.

The key thing is that this result holds for the best estimate of *all* of the averages, not just one of the averages (you will likely be more off with some of the averages, but closer with others). The paradox is even more astounding when you find that the prediction of the National Hockey League's points-per-game average, the average exchange rate for Canadian dollars to US dollars, and the average heights of school-age children can all be predicted more accurately if you incorporate all three, even though they are completely unrelated!

DOES G–D GRADE ON A CURVE?

Sometimes I still want more information, something beyond the numbers. As I mentioned, I love pictures, especially in math. Whenever possible, I draw a picture when I work out a problem. Pictures can be such a vital and beautiful way to capture all of the important aspects of a problem, and a diagram can suggest ways to tackle it.

Histograms, which I talked about in Chapter 3, often tell a more complete story about data than the numbers alone. Data is grouped into ranges of the same size, and a bar is drawn over each range; its height represents the number of numbers that land within the range. A histogram for the test scores would look like the following:

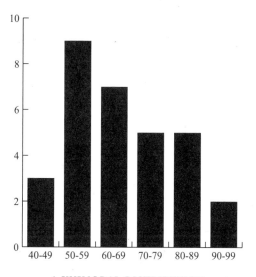

A UNIMODAL DISTRIBUTION

I can see a lot about how the scores are distributed. The distribution is *unimodal*, meaning just a single bar (or consecutive set of bars) is the tallest and hence represents the group of data that is most common. The "uni" in unimodal refers to one mode, i.e., most common value. (Many mathematicians have "unibrows," but I'll talk more later about mathematicians' personal grooming.)

The shape of the curve is what I'd expect, though I have taught classes where the distribution is *bimodal*, that is, there are two distinct peaks among the data, as in the example below. This is indicative of a class that has rather large groups of weak students and strong students, and I know I have my work cut out for me figuring out how to teach the class.

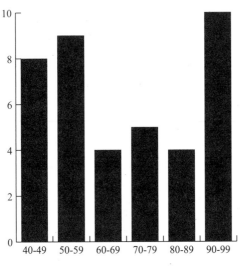

A BIMODAL DISTRIBUTION

I see that the original histogram is not quite the symmetric *bell curve* that is so often talked about; the histogram has a rather long tail out the right. Bell curves are unimodal and balanced about their means, with as many values above the mean as below. In fact, even students who know nothing about mathematics know about bell curves and how the distribution of grades should fall into such a shape, with the mean somewhere around 65%. I'm always disappointed when the first enthusiastic question at the beginning of term in calculus class is inevitably "Do you grade on a curve?" The underlying query is "If we all conspire to party and not study, will you fix the marks so that we do well?" I don't grade on a curve; the grades are what they are. The flipside to moving marks up when they are low is that you have to move grades down when they are too high. Not too many students would accept that bitter pill.

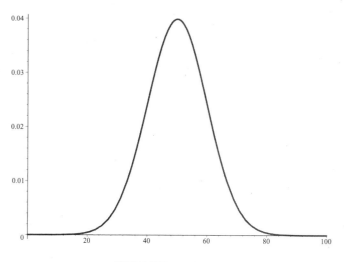

THE NORMAL CURVE

POLLING THE PUBLIC

So much for existential questions for now. Time for dinner! But I do get a little tired of having mealtimes interrupted by pollsters. They always call when I have a mouth full of food – what are the chances? And I am expected to have an opinion on everything. Have you ever noticed that when polls are reported, they always give you a value, along with a "margin of error," and often stating "19 times out of 20." Polls not only reflect public opinion, they also influence it, so it is important to understand the polling process and what the numbers actually mean.

We've talked about things like means and standard deviations; these are numbers that you can calculate, but only if you have access to the whole population in which you're interested. This is unlikely, either because of the size of the population (perhaps all the adults in Canada) or the time and cost involved. What needs to be done is to *sample* only a small group from the population, in the hope that it is representative of the population as a whole. Such a sampling needs to be done randomly, or the sample probably won't be representative. A big part of statistics involves figuring out how to set up experiments with random sampling so that unknown factors that could affect the results are made irrelevant through randomness and design.

This is the problem with all of those online polls. They don't randomly choose people to answer their polling questions – it is purely voluntary, and those who answer likely have a particular bias. Internet polls have the illusion of being fair and reliable, though, and that's why they get under my skin. More bothersome is that the unreliable results of such polls can affect public opinion.

I am looking at the results of an interesting online poll. When 1,000 Americans were asked whether they were in favour of "Canada being annexed to the United States," 38% of those surveyed said yes, while 49% said no. The rest, 13%, did not express an opinion (or asked where the hell Canada was). The polling company, Léger Marketing, stated that the margin of error was "plus or minus 3.1 percentage points, 19 times out of 20." Margins of error vary from poll to poll, but the "19 times out of 20" never changes.

What do these numbers mean? When the poll states that 38% were in favour, it means that out of the 1,000 people surveyed, 38%, or 380, said yes (I guess it could actually be a few more or less, as long as it rounds to 38%). Statistics allows you to generalize from the sample to the whole population. You don't expect the true proportion of people in the whole population who said yes to be exactly the same as the proportion in the sample. Just close. But how close? And how sure do you want to be that you are that close?

Close is usually within a few percentage points; this is the margin of error. So stating that the margin of error is 3.1 percentage points means that the range we are giving for the proportion of people *in the population* that agree with the statement is from 38%−3.1%=34.9% to 38%+3.1%=41.1%. Now the question becomes, how sure are we that the population proportion is in this range?

There is always a chance that the pollsters happened to pick a sample of people of whom more agree or disagree with the proposition than in the general population. It *can* just happen by chance. So the poll numbers can be way off. It should happen only rarely. Now we get down to the crux of the matter. What is rare? Rare shouldn't mean "never" as then in order to say something that is rarely false, we couldn't say anything at all! Rare shouldn't be so large that it includes occurrences that we feel it shouldn't, and on the other hand, it shouldn't be so small that it excludes events that are unlikely.

The consensus among statisticians is that something is rare if it happens only 5% of the time, which is one time out of 20. So our poll needs to be in the range of 34.1% to 41.1% the rest of the time, which is 19 times out of 20. This 19 times out of 20 gives us a comfort zone for how often the poll might be wrong. I don't lose a lot of sleep over things that happen only 5% of the time.

But it's important to keep this 19 times out of 20 in mind for polls. Almost everyone takes what polls say at face value – they're always right, aren't they? But if you took 20 polls, you could expect one of them to be wrong. It would be fun to carry out a poll on how often people think polls are correct. That poll would likely only be correct 19 times out of 20!

POLL DANCES IN THE BEDROOM

I share a common affliction with other mathematicians. I read math books surreptitiously, just for fun. Imagine! Growing up, I had to hide a copy of *Linear Algebra with Matrices* under my mattress so that my brothers wouldn't find it.

Reading about new mathematical ideas can be entertaining, exciting even. I have recently started reading a "stats" book, in which I came across the problem of how to poll individuals about a sensitive matter, about a subject where the "pollee" might be reticent to tell the truth. Polls about what goes on in the bedroom or about taxes are perfect examples. Some people might exaggerate, while others might feel compelled to underestimate. Suppose a pollster wants to estimate what percentage of people cheat on their tax returns. People might not want to admit to doing so. What is a pollster to do? The answer is as follows and it's brilliant.

When polling an individual over the phone, the pollster asks him or her to have a coin ready. The pollster tells the individual to flip the coin. Without revealing the result of the toss, if it comes up heads the person should just say, "Yes, I cheat on my taxes," no matter whether they cheat or not. If on the other hand the coin comes up tails, the person should answer the question truthfully.

Now the pollster has no way of knowing whether a "Yes, I cheat" response is a truthful answer, or just the result of a heads on the toss. But if the pollster samples a large group of people, he or she can estimate quite accurately what percentage of people cheat on their taxes. You would expect the coin to come up heads about half the time, and so you would get a "yes" response for those who tossed heads about 50% of the time. But the remaining percentage of heads would be truthful heads. So, for example, if the pollster found that 68% of the respondents answered yes, then about 18% (68%–50%) out of about 50% truthful responders (who answered when they flipped tails) admit to cheating on their taxes. Now 18% out of 50% is just 36%, and this is the approximate number of people who will admit to cheating on their taxes. Isn't that a clever use of mathematics to get around a sensitive question?

BRAGGING RIGHTS ABOUT P-VALUES

Polls are a measure of the truth, as we believe it, as are statistics. Various medications, for example, have scary side effects and there are major questions about whether any given drug is effective. We often listen in rapt attention when we hear anecdotal evidence about one or the other. But things aren't so simple. One can find different people willing to swear on each side of an issue. Is the drug effective? I can find someone who will confirm it. The medication doesn't work? I can find patients who will swear to that as well. The same is true about side effects.

So you can't base science on anecdotes. What you need is statistics to make sense of the conflicting evidence. If a drug administered to 100 people causes relief of symptoms in 68 cases, is this good evidence in favour of the drug's effectiveness? Or might it be that such an occurrence is just due to chance, as just by chance some people will improve, even without treatment?

Statistics are a part of good scientific practice. Experiments need to be set up, and data gathered and analyzed. Again, we can never be 100% sure of any experimental result; there is always a chance that we are wrong. If the drug works on 68 patients out of 100, it might be that we got lucky, in that most of these patients would have become better all by themselves. Or perhaps these people were particularly resistant to getting better by themselves, and the results of the experiment are therefore all the more impressive. (When my wife and I were trying to have a child, I read that there was statistical evidence that putting your feet up in the air after making love would increase the chances of conceiving. I tried it a number of times and found that it did little good.)

What statistics measures is how likely you were to get the results you did, or even more extreme results, from an experiment, if you make the assumption that the treatments don't work at all. This value is called a *p-value*, and the smaller it is, the stronger the evidence you have in favour that the treatment works. For example, suppose that in a study for a drug's effectiveness, 27 people out of 40 were cured. Without treatment, you might, by past experience, expect a cure rate of only 50% (which would be 20 people). Statistics tells us that in this example the p-value is 0.02, and indicates that if the treatment had no effect, only 2% of the time would you expect, just by chance, 27 or more people of the 40 chosen for the experiment to be cured. Typically a rare event is something that occurs 5% or less of the time, as we have seen, so any p-value less than 0.05 is likely to be given as evidence that whatever treatment has been utilized is indeed working. (That is, if the p-value is less than 0.05, then the chances of observing what you have observed – or something even more extreme – assuming that the treatment has absolutely no effect, is only 5%. As 5% is a relatively small probability, it is very likely the treatment is having some effect.)

All of this is important when I look up medical articles on the Internet. I'm not a hypochondriac, though I often worry that I am. Understanding p-values allows me to at least read the conclusions of these articles with some feeling for the implications.

I can just imagine a bunch of statisticians in a locker room, talking:

> *Statistician 1:* Hey guys, you may not believe it, but my p-value is only 0.04.
> *Statistician 2:* I'm not one to compare, but the last time I measured, my p-value was less than half of that!
> *Statistician 3:* I wish my p-value was that small!

STATISTICS, DAMN STATISTICS

I'm always amazed at how intimidated people are by numbers. People will argue with me on any topic, but just wait until I throw in a number or two. Discussion over. Numbers are powerful and imposing, and those who wield them wield power. As Mary Chapin Carpenter sang in her hit "I Feel Lucky," "The stars might lie but the numbers never do."

But numbers, and in particular, statistics, can and have been used to bend the truth or mislead. I am always on the lookout when numbers are thrown about. And you should be as well.

The media often have frightening stories about incidences of some rare but serious disease that is linked to some treatment. But if a large enough number of people receive a treatment, some will come down with any given disease. For example, if doctors noticed a number of cases of juvenile diabetes among children who have received this year's flu shot, the public would be alarmed. But there are going to be a number of such cases with or without the flu shot, and finding that most of the new cases of juvenile diabetes have received the flu shot in and of itself would not be remarkable, as so many children get flu shots. That is not to say that there isn't a problem, simply that more evidence needs to be gathered before jumping to conclusions. What is more important than the large number of patients who have juvenile diabetes and had the flu shot is the percentage of those who have been diagnosed with juvenile diabetes among those having had the flu shot, as opposed to those diagnosed without having had the flu shot. (Even this comparison could be skewed, as those who are getting a flu shot are visiting their doctor and therefore are more likely to be diagnosed.) Statistics is a tricky thing to get right.

Correlation is the mathematical term for how well two quantities are related. You would expect a person's height and weight to be somewhat correlated – in general, the taller the person, the more he or she is likely to weigh. Level of education and salary would be another example. Quantities can also be correlated, though negatively – increasing one may be associated with a decrease in the other. Level of mathematical education and number of women dated would likely fall into this category.

But many things are not correlated. Height and IQ would be

an example. There is no pattern that you could use to predict the IQ of someone based on their height. Mathematically, correlations range from −1 (negatively correlated) to 1 (positively correlated) with those that have value 0 being uncorrelated.

The news often cites studies where some pair of quantities is found to be highly correlated; statistics is used to quantify the nature of the correlation. The tricky subtlety is that correlation is not the same as causation. Statistical proof that two quantities are related to one another – increases in one quantity tend to be related either to increases or decreases in the other quantity – does not mean that one factor *causes* the other. There could, for example, be something else at work that causes both factors.

Drinking alcohol might be highly correlated with lung cancer, for example, but you shouldn't draw the conclusion that drinking *causes* lung cancer. What may be true is that both drinking and lung cancer are correlated with smoking, a third factor that may be a root cause of lung cancer (though not of drinking).

I have read that there are changes in the structure of the brain that occur until a person reaches approximately 20 years of age, and this fact has often been used as "proof" of why adolescents make poor choices with regard to risky behaviour. But some statisticians have pointed out that the correlation between changes in the structure of the brain and risky behaviour should not be used to draw the conclusion that the former *causes* the latter. It may be true, but it may not be. Scientists may yet find out that something else entirely, such as changes in hormone levels, is linked to both brain development and risky teenage behaviour and may be the true cause of both.

A TRICK THAT WORKS (MOST OF THE TIME)

Thankfully, probability and statistics can be used for good, as well as for evil. Here is one of my favourite party tricks, one that you can try at any large gathering.

The next time you are in a room with at least 23 people, announce that you, with your innate psychic abilities, have determined that there are two people in the room who were born on the same day of the year. Ask everyone in turn to state their birthday. More than likely, someone will stand up and say, "That's my birthday, too!"

Now certainly if there are at least 367 people in the room there will be two people born on the same day of the year. This is an example of what is known as the *pigeonhole principle* – if you place pigeons into pigeonholes and you have more pigeons than pigeonholes, there must be at least one pigeonhole with more than one bird in it. Why mathematicians would be moving pigeons in and out of pigeonholes is another matter (we are a strange lot). Now imagine the days of the year as pigeonholes (there are 365 of these, 366 if you count leap years), and each person is placed into the pigeonhole corresponding to his or her birthday. If we have at least 367 pigeons, er, people, we'd have to have at least two people in the same pigeonhole, that is, with the same birthday. I'll talk more about the pigeonhole principle in Chapter 9.

Unless you have at least 367 people in the room, I don't recommend trying to stuff people into pigeonholes. With fewer than 367 people, you can't be sure that there will be two people with the same birthday. But the probability is on your side as long as there are at least 23 people. Why? Focus in on one particular person, say Mr. A. The chance that another person (Mrs. B) in the room has a different birthday from Mr. A is 364/365, as Mrs. B's birthday must be one of the 364 days of the year different from Mr. A's birthday (for the purposes of this example we'll ignore February 29 as a birthday, as it happens so rarely). The probability that the next person, Miss C, has a different birthday from both Mr. A and Mrs. B, given that they have different birthdays, is 363/365, as her birthday must be one of the 363 days of the year different from Mr. A's and Mrs. B's birthdays. Continuing (and omitting a few details) you find that the probability that everyone has a different birthday is at most $(364/365) \times (363/365) \times (362/365) \times \ldots \times (343/365) \approx 0.49$. That is, at most 49% of the time everyone would have different birthdays. But this means that at least 51% of the time, at least two people would share the same birthday.

This means that more than half the time you play this trick, you'll come out smelling like roses. If there are more people in the room, the probability jumps rather quickly in your favour. For a room with 30 people, the chances of your being correct rises to more than 70%. For a room with 50 people in it, you are 97% certain of being correct. Chances are you'll impress everyone.

The talk on the radio is about baseball records, with much to-do about Barry Bonds eclipsing Hank Aaron's previous record of 755 home runs. When I was young, I'd follow the sports statistics for my favourite players, but now I'm content just to watch a game in passing.

I am still driving around at the university, looking for parking. I find that the probability of finding a parking spot decreases significantly for every minute past nine o'clock. It's 9:13, I'm late yet again, and I am driving through one of the outer parking lots. Eureka! I find one that's really a three-quarter spot. With a bit of spatial geometry I manage to squeeze in. There's that Law of Large Numbers again – even small probability events can come true.

5

SUDOKU, ANYONE?

Regrettably, I don't have a lot of time to play games. Who does? In my undergraduate days, I would while away the hours playing Donkey Kong in the university's games room. Boy, that dates me like Methuselah. When I was much younger, I loved to play Ping Pong and pool, and was lucky enough to grow up in a house that had both types of tables. Friends would come over and we'd play all day. My mother would whip up a bowl of her famous sour cream fudge, which we would polish off while playing board games. Some of my fondest memories are of playing Monopoly with my younger sister and brother, and attacking my baby brother when he got the giggles after winning for the umpteenth time.

People love playing games. As a child, I spent a lot of my time playing games and, like most small children, I had a need to win.

Now whenever a board game is brought out, the expectations on me are always high. Everyone figures that the math professor must have an innate advantage.

I used to play all sorts of games – dots-and-boxes and tic-tac-toe in the car on long trips, board games like Snakes and Ladders, Monopoly, and Stratego, and athletic games like soccer, baseball, and street hockey. I was always looking for an edge in a game, and the more I could strategize, the better. I relied on deductive reasoning, the cornerstone of mathematics, to plan my attacks. A mathematical mindset was very useful.

I remember once having a visit from a relative, Boris, a former Canadian chess champion, back in the 1950s. He and I sat down to play, and he graciously let the game last longer than it should have.

IF ONLY I HAD PERFECT INFORMATION

It wasn't until much later, at university, that I learned that there is a mathematical theory about games, and I was enthralled. I never knew that there was so much that mathematics could say about games in general. Much of the work in this area had been published relatively recently, since the 1940s.

There are some games that have perfect information – that is, each player knows exactly what the rules are, what the other player's moves have been, and what they might be, based on what has transpired. Games of perfect information that end after a finite number of steps (that is, they can't go on forever) and for which there are no draws must have a winning strategy for one player or the other. This is one of those general laws of mathematics I've spoken of. And it illustrates the difference between theory and practice. A game may have a winning strategy for one player or the other, but finding it is another matter!

Let's take a game that you must have played when you were younger, tic-tac-toe. We'll call the players X and O. (Mathematicians always like to name people with single letters if possible – it's probably good that we have spouses to help choose our children's names. Otherwise we'd be tempted to call them A, B, C . . .) A game ends either with one of the players winning, or in a tie, after at most nine plays, starting with X taking the first turn. The rules are known to everyone and all moves are out in the open. Can one of the players force a win? After going through the moves, I can see that a tie is the best both players can hope for, though if either makes a mistake, then the other player might be able to capitalize on it.

Here is how I arrive at this conclusion: For X's first move, there are nine choices but only three different types of first move – playing in a corner, playing in the centre square, and playing in a middle column or row. Notice that there is an inherent *symmetry* about choosing the first square. For example, playing in any one corner is pretty much the same as playing in any other corner; you could rotate the square around the centre by 90 degrees, 180 degrees, or 270 degrees, and continue to play, and the game would be the same. Observing symmetry in problems can help bring huge problems down to much smaller ones. It's one thing that mathematicians are notoriously good at, and much better, in fact, than computers.

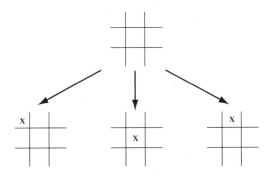

In each of these scenarios there are only a few moves, taking into account symmetry, that O can make. For example, if X's move was to the centre box, then O's response must be one of the following:

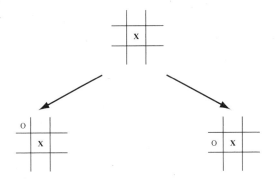

Let's consider first the right diagram, where O responds in a middle row or column. If X then plays in the lower left square, then O must respond in the upper right corner or lose. But then X plays in the lower right square and wins, as he has two ways of forming a line of X's and O can only break up one of these with her next move.

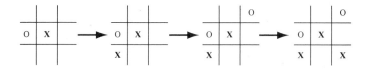

So in this case, provided X has moved to the centre and O has responded in one of the middle rows or columns, X wins. That means, if X moves first in the centre square, O shouldn't respond in one of the middle rows or columns, because then X has a strategy to win.

If instead O had responded by playing in a corner (say the upper left), it is more difficult to determine who has the advantage. But then again, there are only a few ways that X can respond:

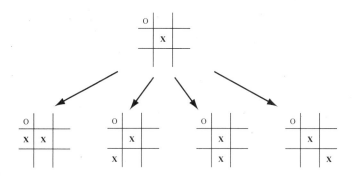

For the first one, O must respond in the square at right in the middle row:

Now we have some choices for how X might play. What he shouldn't do is play in the lower left corner, as O will respond in the upper right corner, resulting in two ways of completing a line, only one of which X can block on his next move. You can also check that if X plays in any of the other squares, O can respond so that the game ends in a draw, and she cannot force a win for herself. Thus the best X can do is force a tie in this case. The same type of reasoning will show that for the game starting with X in the centre and O in the upper left corner, X can, at best, force the other three positions to end in ties as well.

But these are only the cases starting with X in the centre. If you go through all the other cases you will find that in all cases for X's opening move, O can force, at best, a tie. In fact, my arguments show that for the possible opening moves both by X and O, some lead to wins by one player, and some lead to ties:

O · / X	O X	O X /	X / O
Tie	X wins	Tie	X wins

X / O	X / O	X / O	X O
Tie	X wins	X wins	X wins

X O	X / O	X / O	X / O
X wins	Tie	X wins	X wins

If X plays in the centre, O should play in one of the corners. If X plays in a middle row or column (but not in the very centre), O can play either in an adjacent corner or in the centre. And if X plays in a corner, O should respond in the centre. All of these will lead to a tie game, provided everyone plays optimally for themselves.

I hope I haven't ruined the game for you! Fundamentally, mathematics is a way of thinking clearly and of reasoning, as this game demonstrates. All you need is some padding on your bottom: the fortitude to persevere until you exhaust all cases and cover all bases. It's your "sweat equity" in the solution that counts. You wouldn't think of paying someone to work out for you. Why should you forego the pleasure of exercising your mind?

THE GAME PEOPLE PLAY

Here is another two-person game I often play with my children, as it is easy to explain, fun, and the winning strategy can be understood without much math. It is a variant of a game called Nim, and goes as follows. You start with a stack of coins. One of you goes first. Each player is allowed to remove one, two, three, four, or five coins from the stack on his turn. The object of the game is to be the one to remove the last coin.

So, for example, if there are 29 coins present, and I go first, I might take 5 coins away to leave 24 coins. You might take 2 coins away to leave 22 coins, and then I might take away 4 coins away to leave 18. You then might take 5 coins away to leave 13, and I might take 1 away to leave 12. You then might take away 3 coins to leave 9, and I might then remove 3 more coins to leave

6 coins. Now no matter how many coins you might remove, I take the rest and win.

You can play as often as you like, and results may vary, depending on how many coins you start out with and who you are playing. But there must indeed be a winning strategy, a best way to play, as the game has perfect information and ends after a finite number of steps. What is it?

The strategy is best worked out backwards. This approach of working backwards from the end of a problem to the beginning (what I refer to as "bass-ackwards") is one of the most useful strategies in mathematics – and in game playing.

What amount of coins would you like to leave your opponent just before your final move? If you can leave him with 6 coins, you are guaranteed to win – if he takes 1 coin, you take the last 5 coins, if he takes 2 coins, you take the last 4, and if he takes 5 coins, you take the last coin.

Now how can you make sure you leave him with 6 coins at the end? The same type of reasoning will tell you that if you leave him with 12 coins, you can, after both he and you move, leave him with 6 coins by his next move (and you win). To leave him with 12 coins on a move, you need to leave him with 18 coins on his previous move. To leave him with 18 on a move, you need to leave him with 24 on the previous move, and so on. So what you need to do at every move is make sure you leave your opponent with a *multiple of 6* coins on his every move, and you are guaranteed to win.

That is why in our trial game starting with 29 coins I took 5 coins to leave you with 24 coins, and I made sure to leave you with 18, 12, and 6 coins on subsequent moves.

Could you have won? Well, only if you insisted on moving first *and* moving from 29 coins to 24. Then I would have been stuck. Of course, if your opponent doesn't know the winning strategy, you could make any old plays, provided the number of coins is large, realizing that as long as you can switch to leaving multiples of 6 coins at some point, you are bound to win.

I'LL BOX YOU IN

Not all games can be analyzed as easily as tic-tac-toe and Nim. Most often, even in games of perfect information, the best strategies, while theoretically possible, are beyond our abilities, even with computer assistance. Chess is one such game.

Another such game is dots-and-boxes. My younger brother and sister and I used to take turns playing each other on long car rides. To start playing, make a rectangular array of dots:

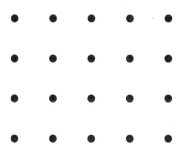

Each player, in turn, joins two adjacent dots, either horizontally or vertically. If you complete a box, you put your initial in it, and you are granted another move. Otherwise, the play turns to your opponent. After no more dots can be joined, the game ends, and the boxes are counted. The player who has marked the most boxes wins.

Let's play. I'll go first.

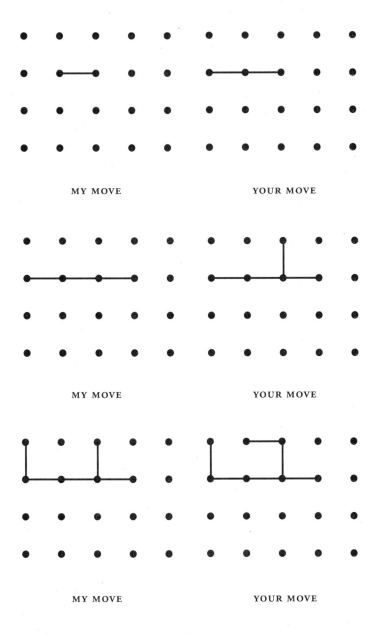

MY MOVE   **YOUR MOVE**

MY MOVE   **YOUR MOVE**

MY MOVE   **YOUR MOVE**

Now I can complete a box, and get another move to complete a second box.

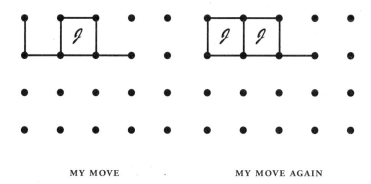

MY MOVE   ·   **MY MOVE AGAIN**

Now there are no more boxes to complete, so I play again somewhere.

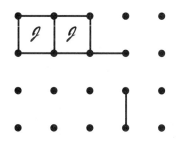

MY MOVE

The play continues until there are no more points to join.

It is clear that one player has a winning strategy. But what
is it? For a large board, there are so many possibilities for play
at each turn that no one has figured it out! But that shouldn't
hold you back. I found a trick in the book *Winning Ways for
Your Mathematical Plays, Volume 3*, by E.R. Berlekamp, J.H.
Conway, and R.K. Guy that can help you win. The secret is
knowing when *not* to complete boxes! This is absolutely
counterintuitive. It seems obvious that you should take a box

whenever it is offered. But usually toward the end of the game the board breaks up into long chains of potential boxes, requiring only that a single edge in a chain be added in order for the next person to take all of the boxes in the chain. Consider the following example (you already have the only filled box, so you are ahead), and suppose it is your move.

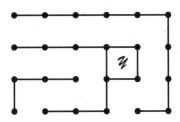

There is a chain of six potential boxes and a chain of eight potential boxes. You would probably add a line in the smaller chain of six (as shown below), yielding it to me, and wait for me to give you the longer chain of eight boxes, so that you would win with 8+1=9 boxes to my six.

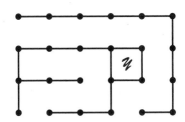

Now I will certainly take you up on your offer of the chain of six, but I'll show some restraint. I'll take four of the six boxes, as shown, but hold myself back.

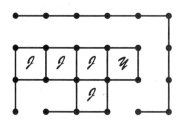

I'll deliberately give you the last pair of boxes in "my" chain by adding the bottom edge to give you a "double box," one that you can divide into two boxes, to fill in with a single line. What a gift!

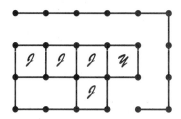

You instantly take the present, and fill in two squares for yourself.

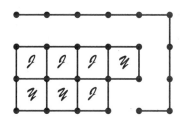

But now you are stuck. Any line you fill in will give me the whole chain of eight, so I will win, 12 boxes to three.

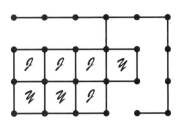

YOUR BEWILDERED MOVE

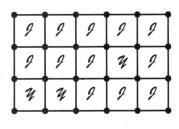

I WIN!

The moral of the story is that it is sometimes invaluable to consider what assumptions you are implicitly making, and whether they are necessary. Here the assumption is that when you are trying to win, it is always in your best interest to grab as much as you can whenever you can. But sometimes it is better to give a little to gain a lot.

OFF THE BEATEN PATH

Sudoku is a recently popularized puzzle that centres on making deductions. In Sudoku, you have a partially filled 9×9 array of numbers. The object is to fill in the missing numbers so that each of the numbers 1 through 9 appears once in every row and column, and once in each of the nine 3×3 arrays that divide up the original 9×9 array (these are shown with heavy lines below). So a typical Sudoku might look something like the following:

9	4							
			8	9			3	
2	5	3				6	9	
	2		9	5		3	7	
	7			6			5	
	1	6		2	7		4	
	3	1				7	6	5
	8			4	6			
							8	4

Ironically, Sudoku has nothing to do with numbers themselves; any set of nine different symbols could be used instead of the numbers 1 through 9. It's just that the numbers 1 through 9 are easy to draw and used by most cultures. But it does give the illusion that numbers are important, so I give Sudoku full points for that.

While I don't need to use any facts about numbers to solve the puzzle, the systematic, logical approach of mathematics is exactly what is required. Of course, rows, columns, and 3×3 boxes that are close to being full are easier to fill in than those that only have a few numbers filled in. I realize that the second column needs a 6 and a 9, and as the second row already has a 9 in it, the six must go in that row.

9	4							
	6		8	9			3	
2	5	3				6	9	
	2		9	5		3	7	
	7			6			5	
	1	6		2	7		4	
	3	1				7	6	5
	8			4	6			
	9						8	4

I can now fill in the upper left 3×3 box using the same reasoning.

9	4	8						
1	6	7	8	9			3	
2	5	3				6	9	
	2		9	5		3	7	
	7			6			5	
	1	6		2	7		4	
	3	1				7	6	5
	8			4	6			
	9						8	4

Most often, I get to a place where I need to consider possibilities and cases. I find it convenient to write lightly into the cells the choices that remain for that cell. A good part of mathematics is finding the right way to visualize and condense the many choices and cases. Here is where I'm at right now:

9	4	8						
1	6	7	8	9			3	
2	5	3				6	9	
8	2	4	9	5	1	3	7	6
3	7	9	4	6	8		5	
5	1	6	3	2	7		4	
4	3	1	2	8	9	7	6	5
	8			4	6			
	9						8	4

For the last two empty boxes in row 5, I have a choice of 1 and 2, and I can't decide which should be in which box. So I'll pencil in both numbers lightly. Similarly, the empty boxes in the next row are 8 and 9 in some order. Here is a general principle that I repeatedly apply: I hunt down boxes with one number that appears only once in its row, column, or 3×3 box; such a number is forced into the cell, and the choices for other empty boxes in the same row, column, or 3×3 box can be updated.

9	4	8						
1	6	7	8	9			3	
2	5	3				6	9	
8	2	4	9	5	1	3	7	6
3	7	9	4	6	8	1 2	5	1 2
5	1	6	3	2	7	8 9	4	8 9
4	3	1	2	8	9	7	6	5
	8			4	6			
	9						8	4

The numbers 2, 4, and 5 must go in the empty spots in row 2, and as the last column has a 4 and a 5 already, a 2 must go in the last box in row 2. I can then update the boxes. It doesn't take too long before I fill in the whole puzzle.

9	4	8	6	3	2	5	1	7
1	6	7	8	9	5	4	3	2
2	5	3	7	1	4	6	9	8
8	2	4	9	5	1	3	7	6
3	7	9	4	6	8	2	5	1
5	1	6	3	2	7	8	4	9
4	3	1	2	8	9	7	6	5
7	8	5	1	4	6	9	2	3
6	9	2	5	7	3	1	8	4

IT'S A-MAZE-ING

My children have always enjoyed doing mazes. All mazes can be solved pretty quickly, if you know the right way to look at them. Knowing what to ignore and what to keep is one of the fundamentals of mathematics.

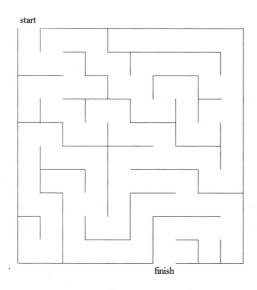

start

finish

As a mathematician, I see that the lengths of the various corridors and their twists and turns are irrelevant to solving the maze. What is important are the places where the corridors meet, and which corridors lead to which other ones. I highlight these by placing a dot at the start, at the finish, and at each intersection of two or more corridors, and I label each for future reference. I ignore any corridors that lead to a dead end, as I would never travel down such a corridor (unlike in real life, where my wife and I routinely get lost while driving).

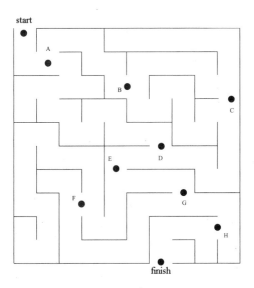

I join two of the dots if a corridor connects them. This encodes all of the important information I need to solve the maze, and I am free to ignore the rest.

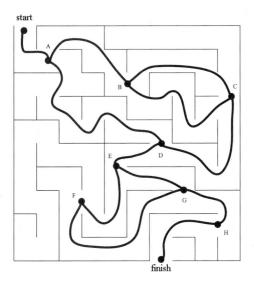

Now the picture seems clearer. What I am looking for is a way to walk from the point labelled "start" to the one labelled "finish." But now, by hiding the inessential and highlighting the essential, I can see a number of solutions, such as the one shown below.

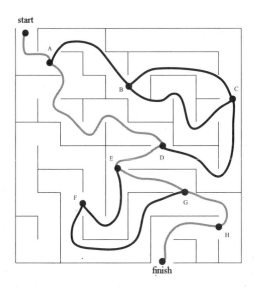

In fact, the easiest way to see the solution is to begin at the point labelled "start" and walk simultaneously to all points you can reach by a new line, labelling a new point by the previous point you used to reach it. Then once you hit the "finish" point, you can track back through the points. The finish point was labelled "from H," so I walk back to H. Then H was labelled "from G," so I walk back to G. Continuing in this way, I quickly get the path shown above. This is a procedure, or an *algorithm,* for finding a path from one point to another in a graph, and a method of solving mazes.

THE LESSONS IN GAMES

If you have children, I recommend spending lots of time playing games with them. There is so much you can teach through game playing – how to strategize, how to focus on the essentials, how to manage chance – and there is a lot of mathematical reasoning that they can pick up while trying to win.

One of my boys' favourites is Mastermind. In this game, players take turns guessing a hidden code, marked by four coloured pegs in the holes in the board. The "encoder" tells everyone how many pegs have the right colour and are in the right spots, and how many pegs have the right colour but are in the wrong spots. For example, if the encoder has set the pegs, from left to right, to be Yellow, Yellow, Green, Red, and a "decoder" has guessed Pink, Yellow, Yellow, Green, then the encoder states that there is one peg of the right colour, in the right spot, and two pegs that are the right colour but the wrong spot. Each guess and the resulting information is recorded on the game board. The object of the game is to break the code, and only 12 guesses are allowed in total. If no one guesses the code, the encoder wins.

Of course, mathematicians have analyzed the game and come up with a variety of strategies to break the code quickly, with the best guarantee being five moves or fewer. What is interesting is that the strategy is not so easy to describe or work out in practice – you'd need a computer to help you make your decisions along the way.

WHO'S A JUGHEAD?

Before mathematics became my profession, I used to enjoy reading puzzle books (now it's too much like work!). You probably remember the puzzle about a boat crossing a river with a goat, a wolf, and a head of cabbage on board without anyone or anything being eaten. Another famous one involved trying to find a way to measure out a certain amount of liquid (say, 7 ounces) using only two jugs of different sizes (say, 5 ounces and 11 ounces) and an endless supply of water. (You can tell how old the puzzle is as it uses ounces instead of millilitres.) The best strategy is to work backwards just as we did for the game Nim. If I can arrange to have 1 ounce in the 5-ounce jug, then I can fill the 11-ounce jug, and fill the 5-ounce jug from it. That will pour off 4 ounces from the 11-ounce jug, leaving the desired 7 ounces in the jug.

So now I have reduced the problem to getting 1 ounce into the 5-ounce jug. But this is pretty easy to do. All I need to do is fill the 11-ounce jug, and pour it off twice into the 5-ounce jug (emptying the 5-ounce jug after each time), to get 1 ounce in the 11-ounce jug. I can then transfer it to the 5-ounce jug, and I am where I want to be.

Mathematicians are particularly adept at converting a complex question into another, simpler one. This reminds me of the first math joke I ever heard. What does a mathematician do after he lights a match? Answer: He blows it out. Now what does a mathematician do if his house is on fire? Answer: He lights a match and reduces it to a previously solved problem.

It's just after noon, and I'm sitting down in the faculty lounge having my lunch. There is a newspaper open to the crossword puzzle, where, as usual, a number of people have left their mark trying to solve it.

I enjoy wordplay, though my wife and I find as we get older that we sometimes have to hunt for words. Just last week my wife was in a tizzy trying to remember the word for that disease where you start forgetting (10 letters, beginning with an "A"). Oh, oh – that reminds me! I have to go on a quick jaunt out to the airport. No time for game playing now; I'm on a mission.

6

CHANCE, DECISIONS, AND THE FEAR FACTOR

It's a scenic drive out to the Halifax airport. My wife is on her way back from visiting her family in Winnipeg. I remember my first trip there, as the groom apparent. Extended family is very important to my wife, and the visit left me with a better grasp of infinity.

City planners have ensured that the airport is located well outside the city, so I need to leave about 30 minutes to make the trip. My mind wanders on drives, especially if I have a research problem that I'm working on. Network diagrams alternate with the road signs along the way.

As I park, I realize that I'm early and I wonder whether I have been speeding or not. I can't recall looking at the speedometer. I remember that the *Mean Value Theorem* in calculus, one of those theoretical but incredibly useful mathematical results, tells me that at some point along the way, my actual speed was at least my average speed over the whole trip. It's a guarantee – don't know where or when, but it's certain. I did set the trip odometer when I hit the highway. I take a quick reading; it reads 33 kilometres, and my watch tells me that the trip took only 18 minutes. Now 18 minutes is 18/60ths of an hour, that is, 3/10ths of an hour. So my average speed on the drive out was the total distance, 33 kilometres, divided by the time it took, 3/10ths of an hour, that is, $\frac{33}{3/10}$ = 110 kilometres per hour, a little too fast for the 100 km/hr speed limit. Next time I had better watch my speed.

I am now waiting for my wife's flight, which the arrivals board says is delayed. The weather has been bad in Winnipeg, but when isn't it? The winters are so cold that it doesn't matter whether you measure the temperature in Celsius or Fahrenheit, because –40 degrees is the same in either scale.

It's been a while, and still no sign of her flight. I'm getting a little nervous, and I think back to a flight I took several years earlier to attend a mathematics workshop ...

It had been a long, long day of travel, longer than a mathematician's nose hairs. I was en route to a conference at Rutgers University in New Jersey, and I was delayed at Pearson International Airport in Toronto. The weather had turned ugly, with big, math-hating storm clouds outside. As I stood there contemplating nature's fury, I heard over the airport loudspeaker that flights to the eastern seaboard of the United States were being cancelled on account of tornados. Tornados! I had never experienced a tornado, and the word brought to mind Dorothy, Toto, and a small house falling from the sky.

When I looked at the long list of cancelled flights on the departures board, I noticed my flight was the only one to the east coast that was still scheduled. As I tried to decide whether or not to get on the plane, I began to panic. I looked around and saw that there were several other mathematicians on my flight (we may possess a lot of skills, but dressing in front of a mirror is not one of them). I thought to myself, "What are the chances – or probability – that a plane with several mathematicians on board would crash in a storm?" Satisfied that a crash was highly unlikely, I boarded the plane, but not without saying a prayer. A couple of hours later, we arrived safely in Newark.

My reasoning was highly questionable. Desperately looking for a way to set my mind at ease, I had allowed myself to fall for an age-old probability trap known as the *gambler's fallacy*. We all believe that the law of averages tells us that if an unbiased coin has come up heads 10 times in a row that the next toss has to be tails, because the odds of tossing heads 11 times in a row are very small. However, the probability is still only 50–50: tossing 10 heads in a row in no way affects what happens on the next toss. The same holds true if you are sitting around the poker table. Even if you've had poor luck all night, you are no more likely to win the next hand than if you had won all of the previous hands. The chances do not change based on what has happened previously, no matter how unlikely the events leading up to the present.

Nevertheless, I still find it extremely difficult not to fall for the gambler's fallacy that an extraordinary run of bad or good luck has got to change, and change soon. This conclusion hits me on an emotional level – I feel it in my gut. And while it is true that it is extremely unlikely there will be a group of mathematicians on

OUR DAYS ARE NUMBERED

a plane that crashes, it does not follow that the presence of several mathematically like-minded souls will confer additional protection.

While it is true that the probability of a plane a) having many mathematicians on board and b) crashing is extraordinarily small (I would say at most one in 100 million), it is not relevant to the plane's safety. The small likelihood that a plane bearing a number of mathematicians will crash has as much to do with the small number of mathematicians versus the number of people who fly, and the likelihood of those mathematicians sharing a flight, as it does with the likelihood of a crash. All that matters to my decision is the probability of a plane crashing, which, though still small (roughly one in a million), is undoubtedly higher than that of a plane crashing while carrying a number of mathematicians. To be fair, what really matters is the probability of a plane crash given the existing conditions, such as the weather. Even if the poor weather increases the probability of a crash by a factor of 100, say to one in 10,000, I shouldn't worry too much. Pilot error is a much bigger worry. If I were to see the pilot and copilot board the plane with beer bottles in hand singing drinking songs, I'd certainly revise my level of risk!

Reminiscing about my decision of whether to fly or not makes me think about decision making – not about how we make decisions, but rather how we *ought* to make them. Is there a rational way to choose among our options? Is there a better way to make decisions, or are we left to make them by the seat of our pants? There is often a great deal at stake when we need to navigate through a river of problems, so it's not all that surprising that mathematicians have waded in to help.

CRIKEY!

Making a single risky choice isn't necessarily a bad idea, but repeated dangerous choices tempt fate. I remember the day naturalist Steve Irwin died; my younger son cried himself to sleep that night. The great crocodile hunter was brought down by a freak event: a stingray's barb pierced his heart. How did such an unlikely event happen? It seemed shocking that someone so accustomed to the dangers of wildlife should die this way.

A quick calculation was enlightening. Steve Irwin made a livelihood out of placing himself in dangerous situations. Suppose that on any given day out exploring nature he, with his extensive skill and experience, had only a one in 10,000 chance of something going terribly wrong and resulting in death. The probability of his surviving all animal encounters that day would be 9,999/10,000=0.9999, or 99.99% – pretty much for certain. Over 20 years, or approximately 20×365.25=7,305 days of taking such risks, the chance of surviving all animal encounters would be the number you get by multiplying the probabilities of surviving each day:

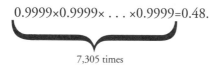

$$0.9999\times0.9999\times \ldots \times0.9999=0.48.$$
7,305 times

That is, there would be a 52% chance of dying in some animal-related event. (Steve Irwin's probability of dying within the 20 years of course would be even higher if you took into account all possible ways of dying.) Now we come up against what seems to be a paradox along the lines of the gambler's fallacy. It was indeed very unlikely (only a one in 10,000 chance) that Irwin would die on that day in the particular way he did. But it was very likely (more than a 50–50 chance) that he would have died on *some* day in those 20 years in *some* animal-related accident. Think about it: if he had died after being trampled by a rhino, or after contracting a virulent infection from a lizard bite, his fans would have been just as shocked and saddened.

We may not like to accept it, but we do get an automatic upgrade to a higher level of risk when we repeat any risky behaviour. So should we limit our travel rather than repeatedly expose ourselves to the risk of a plane crash? It's an excellent question. Undoubtedly, we do upgrade our risk the more we fly, but the important question is to what level. Air travel is so safe that we'd have to fly more than 2.5 *million* miles before the probability of a crash would exceed 50%. And each individual flight would remain as unlikely to crash as ever.

What the calculation highlights as well is the fact that while any one rare event is almost certain not to occur, *some* rare event is eventually going to happen to you – there are no two ways about it. Rare events happen to everyone. Most often the rare events that do occur are not as catastrophic as what happened to Steve Irwin. The rare event may be winning a prize in a lottery, or running into a friend you haven't seen in 30 years while vacationing outside the country.

I do find that knowing that, mathematically, some rare events will happen to me allows me to put such events in perspective. Simply knowing that, just by chance, there will be streaks of bad luck (and streaks of good luck) lets me take them more in stride.

Having said all this, Steve Irwin's death is no less tragic, just perhaps less surprising. He lived the life he wanted to live. One lesson we can take away is that while occasional risks are safe to take, repeated risk taking is an accident waiting to happen. So take a moment and think about the risks you take, how often you take them, and for how long. Is it worth it? Only you can answer that question.

THE FEAR FACTOR

Should we be afraid? Is being afraid a waste of energy or a necessary survival skill? I am a very good worrier, but sometimes mathematical thought puts things into perspective.

Many of us will have at least one medical scare in our lifetime. When I was younger, I found an enlarged lymph node in my neck and went to a walk-in clinic. The doctor examined the node and said, "We'll need to run some tests. It could be lymphoma." Then he sent me home, to spend a weekend sick with worry. Lymphoma! Why me? The bigger question became how to calm myself down. It seemed to me, even in my heightened state of anxiety, that lymphoma was unlikely.

But is there a way to evaluate how frightened I should have been? This involves measuring not the chances that I had lymphoma (which were quite small) but rather the probability that I had lymphoma given that I had an enlarged lymph node. It is an example of what we call a *conditional probability*, as it is based on some condition holding. Most of the probabilities we encounter on a daily basis are conditional probabilities, which depend on the circumstances in which we find ourselves.

To begin, I needed to estimate certain probabilities. From a quick search on the Internet, I found the incidence of lymphoma in the United States for my age group was approximately 2.9 cases per 100,000. Frequencies are really just probabilities in disguise. I took the probability of having lymphoma to be about $2.9/100,000 = 0.000029$, which is 0.0029% (pretty darn low). The probability of *not* having lymphoma was therefore $1 - 0.000029$, which is 0.999971, or 99.9971% (after all, it is rare to have lymphoma).

Now I had to estimate some other probabilities. First, what about the probability that I would have an enlarged lymph node, given I had lymphoma? I would estimate this as pretty high, say 95%. Next, what about the probability that I had an enlarged lymph node, given I didn't have lymphoma? This would be lower, for sure, but not too low – perhaps around 5% – given that an infection could be the cause. These were the important probabilities, namely having my symptom whether there was an underlying lymphoma present or not. I drew a bar graph showing the probabilities of having lymphoma or not, and divided up each according to the probabilities of having an enlarged lymph node or not. The heights of each bar correspond to the associated probabilities, so that the bar on the left has a height of 0.000029, and the bar on the right 0.999971. I've had to significantly zoom in on the tiny lymphoma bar in the top diagram to see its breakdown; this will be crucial in what follows.

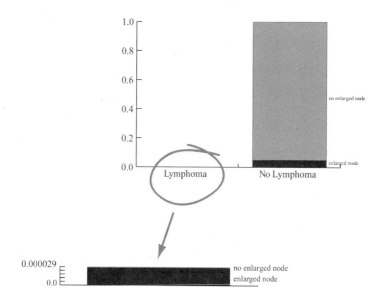

BIG BARS, LITTLE BARS, AND BAYES' THEOREM

How do we put all of this together to find what we are after, namely the probability that I had lymphoma given that I had an enlarged lymph node? Some general mathematical rules can be of great service. There is a well-known result called *Bayes' Theorem* that switches around the order for conditional probabilities. This result dates back to eighteenth-century England and the work of Reverend Thomas Bayes (not so long ago a lot of non-professional mathematicians discovered some of the best mathematics). Bayes' theorem is usually written out as a long equation, but the story is in the picture.

In the bar graph, having an enlarged lymph node is represented by the two black rectangles in each of the two bars. The probability of having lymphoma, given I had an enlarged lymph node, corresponds to the ratio of the height of the black rectangle from the Lymphoma bar to the combined height of the two black rectangles. Bayes' theorem is exactly what you need to calculate this proportion.

You can see from the bar graph that the black rectangle from the Lymphoma bar is insignificant compared to the black rectangle in the No Lymphoma bar, but let's see just how insignificant it is. The height of the black rectangle from the Lymphoma bar is 95% of 0.000029. In Chapter 2 we saw how to convert a percentage to a number, and that "of" translates mathematically into multiplication. Thus this height is 0.95×0.000029, which is 0.00002755. The height of the black rectangle in the No Lymphoma bar is 5% of 0.999971, which is 0.04999855; it is *much* bigger (almost 2,000 times bigger)! So the probability of having lymphoma given I had an enlarged node corresponds to the ratio of the height of the black rectangle from the Lymphoma bar (0.00002755) to the combined height of the black rectangles (0.00002755+0.04999855=0.05002610). That proportion is 0.00002755/0.05002610, which is approximately 0.00055, or just 0.055%. The calculation based on Bayes' theorem shows that the probability I had lymphoma, given I had an enlarged node, was indeed minuscule, only approximately 0.055% – much less than 1%. In the future, I think I'll use Bayes' theorem to temper anxiety resulting from any unconfirmed diagnoses.

The enlarged lymph node turned out to be just that, enlarged, probably due to a viral infection. I never visited that clinic again.

I had a right to be angry with the doctor for even suggesting I had cancer. Why did he even mention it? Either he had such a poor bedside manner that he didn't care if he worried me to death, or – and I take this as the more likely scenario – he didn't understand the difference between the probability of having an enlarged lymph node due to lymphoma, which is high but irrelevant, and the more important statistic, namely the small probability of having lymphoma given an enlarged node. We should all be aware of how easy it is to mix up the order of conditional probabilities, and how this can lead to erroneous conclusions.

ENJOYING LOTTERIES

Of course, there is more to making decisions than understanding probabilities. For example, I sometimes enjoy playing the lottery – definitely not something to do based on the likelihood of winning. In one lottery I choose six numbers from 1 to 49, and get a big payout (which depends on how many people play, but is always in the millions) if I match all six of the numbers correctly. The chance of matching all of the numbers is only one in 13,983,816, an appallingly low chance of winning. There are additional prizes, based on how many numbers you match and whether you also match an additional "bonus" number. Here are the winning amounts and the probabilities of winning for a recent lottery:

MATCH	PROBABILITY	PRIZE (DOLLARS)
6 of 6	1/13,983,816 (approximately 1 in 14,000,000)	3,857,506.00
5 of 6 + bonus	1/2,330,636 (approximately 1 in 2,000,000)	54,841.50
5 of 6	3/166,474 (approximately 1 in 50,000)	1,986.30
4 of 6	645/665,896 (approximately 1 in 1,000)	54.30
3 of 6	8,815/499,422 (approximately 1 in 50)	10.00
2 of 6 + bonus	1,025/83,237 (approximately 1 in 80)	5.00

Like most mathematicians, I like to visualize numbers whenever possible, so I created a bar graph of the probabilities for each prize. Note how tiny the probabilities are for the large prizes in comparison to the smaller prizes – you can't even see the rectangles!

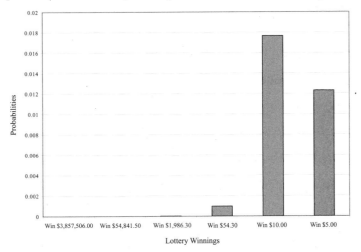

MY CHANCES OF WINNING THE LOTTERY

My overall chance of winning is the number you get by adding up all of the probabilities, here about 0.03, that is, only about 3%. My probability of winning one of the two big payouts is only one in approximately 2 million. I have a better chance of being murdered on a given day, but I wouldn't want to run a lottery founded on that concept.

What can the lottery player expect to win on average? To calculate this, multiply each prize amount by the chance of winning that amount:

$$(3,857,506.00 \times \frac{1}{13,983,816}) + (54,841.50 \times \frac{1}{2,330,636}) +$$

$$(1,986.30 \times \frac{3}{166,474}) + (54.30 \times \frac{645}{665,896}) + (10.00 \times \frac{8,815}{499,422})$$

$$+ (5.00 \times \frac{1,025}{83,237}) = 0.63$$

That is, for every $2 ticket you buy you would earn, on average, approximately 63 cents. You can see why the lottery corporation is doing very well!

So does it make any sense to play the lottery? Well, in fact, yes, and the reasoning centres not only on how we evaluate risk, but on how we assign value to things (or fantasies). Not everyone values the same things in the same way, even something as concrete as money. At this very moment, does $10,000 mean a lot or a little to you? If you desperately need a medical procedure that costs $10,000, it means a lot, but if the same procedure costs $100,000, it means less. The key thing is that $10,000 may not be worth 10,000 times what one dollar is worth to you. Mathe-maticians talk about a person's *utility function*, which describes how much value we place on an outcome. If the top prizes of the lottery are worth a lot more to you than the actual dollar amounts (perhaps they mean instant, permanent, happy retirement, which may seem almost priceless), then it may indeed make sense to purchase a ticket. You are buying not only into a lottery, but also into a dream. And for my biggest dreams, I am willing to go out on a limb and buy a lottery ticket.

Deciding whether to put money in a parking meter is much like a lottery. How much money you decide to put in depends on many things – how long you think you'll be, what the chances are of getting a ticket when the meter runs out, and how negatively you weigh getting a ticket (the fine, the wasted time, and the aggravation that will ensue).

We also use utilities when we decide on an investment strat-egy. It would be simpler if all we wanted to do was to maximize how much money we could expect to earn. But investment dealers look beyond this, realizing that not everyone treats the same amount of money in the same way. They ask you specific questions to gauge not only how you view risk, but also how you view money. They are trying to figure out your personal utility function for your investments.

IF YOU SHOW ME YOUR UTILITY FUNCTION, I'LL SHOW YOU MINE

Is there a systematic way you can determine your utility func-tion? Moreover does everyone have a consistent utility function?

Mathematical decision theory accepts that each person has his or her own unique way of valuing any given object or event. How I assign numbers (or utilities) may seem completely arbitrary to you, and indeed I may not have any scheme to assign these numbers other than a gut feeling. Utilities may change over time depending on a person's values, financial situation, and so on. For some people, the way they value items is so flawed that careful observers can take advantage of them. For example, to assign utilities, your preferences among items must be *transitive* – if you prefer A to B and B to C then you must prefer A to C. It may seem natural and logical that preferences should be transitive, but in practice it seems that many rational people's preferences aren't transitive after all. People whose preferences are not transitive are known as "money pumps," because it is not hard to empty their wallets.

Suppose you prefer a PDA to an MP3 player and an MP3 player to a cell phone, but actually prefer a cell phone to a PDA. Then suppose you own a PDA. If you give me $10, I'll trade you a cell phone for your PDA. For another $10, I'll trade you an MP3 player for your cell phone. Finally, as you prefer a cell phone to a PDA, I'll trade you back your original PDA for the cell phone plus an additional $10. What happened is that because of the lack of transitivity in your preferences, you have given up $30, all for nothing! Just give another crank to the money pump, and the money keeps pouring in!

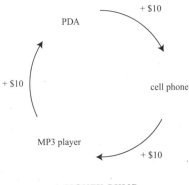

A MONEY PUMP

Other factors can interfere with creating a utility scale. The value of an item may depend on the circumstances surrounding it. For example, can you give a value to what $100 is worth to you? You would probably travel right across town to save $100 on an item worth $110, but probably wouldn't do so for an item worth $10,000. It's the same $100, but what it's worth to you varies.

Utilities also seem to explain in part why young people (I have suddenly found myself outside this group!) engage in risky behaviour, whether it be drug use or unprotected sex. It isn't that they minimize the actual risk (studies have found they actually inflate the risk involved) but rather that they overvalue (in my opinion) engaging in the risky behaviour. The utility they assign to it is much too large compared with the "disutility" of having a life-altering negative consequence. For example, they may value taking drugs and fitting in with the "in crowd" very highly and not rate the medical and judicial consequences of their actions negatively enough. So what we need to talk to our children about is not so much probabilities, but our values and how we value them.

A TALE OF TWO CHOICES

Thinking about risks and your utility function can help focus your mind when making decisions. Here's an example. Suppose you have to decide between coming home for a romantic dinner that will lead to you-know-what or taking an out-of-town client to a restaurant to clinch a business deal. The choices are clear. The outcomes are also pretty clear. Disappoint your spouse and the evening leads to you sleeping you-know-where, and brushing off your client may cause the deal to dissolve to nothing. All that is left to ask is, how do you rank the outcomes? Which is more important to you?

But there is more to the story. While you may estimate that there is an 85% chance of closing the business deal over dinner tonight, perhaps there is only a 50% chance the business deal will fall through if you postpone taking the client out. (Rather than being annoyed, it may well turn out that the client misses their spouse and appreciates your family devotion, so you may be able to close the deal later.)

Attaching probabilities to outcomes is crucial to making decisions. You may base your estimates of how likely an outcome is on past experience, or on data that has been collected. Perhaps you can recall 10 previous instances where you missed a planned dinner alone with your spouse and only one of those evenings was salvageable. This would lead you to estimate that if you choose the business meeting over dinner with your spouse, the homecoming will be sweet only 10% of the time. So you can revise your estimate for missing you-know-what to 90%. More data means more accuracy, so if you are fortunate enough to recall 20 previous instances where you missed a planned dinner (poor you!), you'll have a better handle on your chances. You are 95% sure if you come home for dinner you'll have an evening to remember. We can draw a "tree of knowledge" (or *decision tree*) for your choices and possible outcomes, showing the probabilities on the branches, using "BD" as "business deal." The diamond indicates where you have a choice; the circles indicate where chance plays a role:

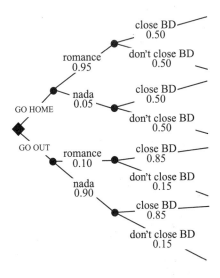

DECISION TREE

WEIGHTY DECISIONS

There is more to decision making than attaching probabilities to outcomes. You need to weigh outcomes, and decide not only which outcomes you prefer, but by how much. Why? We often have to choose not between an obvious good choice and an obvious bad one, but among a variety of options, some good, some not so good. The important thing is not the specific weight we attach to individual choices, but how the various weights relate to one another. The weights we attach are positive if outcomes are essentially good, and negative if they are essentially bad. What we need to do is build a utility function for the value of each outcome.

Back to your evening dilemma. On a scale of 1 to 100, let's weigh each of the outcomes. (Your assignment of weights or utilities is purely subjective, and other people, given the same outcomes, might weigh them quite differently.) After careful consideration, you decide to give a weight of 70 to a great end of the evening but missing out on the business deal, and a weight of 100 to romance plus landing the business deal (it doesn't get much better than that). Clinching the deal with the client, but ending with a night on the couch, you might weight as 50, and missing out on both a romantic evening and the business deal would rate a –10.

DECISIONS, DECISIONS

How should you put everything together? For each of the two choices (go home for dinner, meet with the client), you have weighted your outcomes and have come up with probabilities of these occurring. To find the probability of a branch in the decision tree, multiply the probabilities on the branch. For example, for the choice "Going Home for a Romantic Dinner," the probability of "You-know-what and close the business deal" corresponds to the top branch in the tree: go home → romance → close BD. The two probabilities along this branch are 0.95

and 0.50. The resulting probability of both romance and closing the business deal is the number you get by multiplying these two numbers: 0.95×0.50=0.475. All of the other numbers in the tables below can be calculated in the same way.

CHOICE I: GOING HOME FOR A ROMANTIC DINNER

OUTCOME	UTILITY	PROBABILITY
You-know-what and close business deal	100	0.475
You-know-what and don't close business deal	70	0.475
Miss out on you-know-what but close the business deal	50	0.025
Miss out on you-know-what and don't close the business deal	−10	0.025

CHOICE II: TAKING CLIENT OUT TO DINNER

OUTCOME	UTILITY	PROBABILITY
You-know-what and close business deal	100	0.085
You-know-what and don't close business deal	70	0.015
Miss out on you-know-what but close the business deal	50	0.765
Miss out on you-know-what and don't close the business deal	−10	0.135

How should you evaluate your choices? We can compare the *expected* utility of each choice – we multiply each outcome by its probability of occurring and add the results. In this way we balance off the value you attach to each outcome with its probability of actually happening. This is one common mathematical way to define rationality in decision making. Here is what we get for the example:

Expected utility of Choice I: 81.75
Expected utility of Choice II: 46.45

Either way, you get dinner, but based on expected utility, you would choose Choice I, going home for the romantic dinner, over Choice II, taking the client out to dinner.

CHOOSING THE VERY BEST

Expected utility isn't the only math that is useful in decision making. My wife and I are thinking of selling our house. So here is the problem I imagine that we might face. Suppose we know that we are going to receive, say, eight different offers on our house within the next two weeks. Each will probably come in on a different day, via different real estate agents, and we will be given only a day to respond. Once we turn down an offer, it is off the table. The question is, which offer should we decide to take? We have no idea what amount each offer will come in at (and those who are offering do not converse with one another), or in what order – whether the largest offer will be first, last, or somewhere in the middle. We assume the offers will be in a random order and that no two will be equal.

One option is to decide to pick an offer in a fixed position, say the second offer we get. The chance of that being the biggest offer is only 1/8, or 12.5%, as the largest offer could equally likely be any one of the eight offers. Can we do better?

We can. We can decide to let a certain number, say r, of the offers go by, and then take the next offer that is bigger than all the first r offers. For example, we can decide to let the first two offers go by ($r=2$) and pick the next offer that is bigger than the first two offers (we may be forced to choose the last offer if there are no offers better than the first two).

Mathematically we can show there is a best choice for r that makes the chances of picking the largest offer the greatest. For eight offers, what you should do is let three offers go by ($r=3$) and then choose the next offer greater than those three (or failing that, the last offer). Your chances of picking the best offer go up to about 33%, one in three, rather than one in eight (12.5%).

This problem is an example of what is known as the *Secretary Problem*, or the *Sultan's Dowry Problem*. (Problems in mathematics often go by several names, depending on who named them and the underlying real or imaginary problem that evoked them.) In all such cases, what you have is a given number of items you can rank from best to worst, which pass by you one at a time, in some unknown, random order. You want to choose the best, but once you decide to pass on an item, you can't go back. What is fascinating is that if the number of items is large, then the best option, mathematics tells us, is to let approximately 37% of the numbers go by, and then pick the next item that is larger than the largest of the first 37% of the items (or the last item, if you come across no such item).

I think the reason some people never pick a spouse is that they have no strategy for picking the best among those they date. They are worried that they have already passed on the best, or that the best is always in the offing, so they never make a choice. I can just hear George and Jerry talking in the coffee shop:

George: We're never going to get married and do you know why?

Jerry: No, but I think you're going to tell me.

George: The problem is we never know when we've met Ms. Right. There might always be someone better.

Jerry: I know they must be thinking that.

George: Well, you know what? Now I'm changing that. Here's my strategy. I'm going to marry the best of the next 20 women I date. After dating 37% of them, I'll pick the next one who's better. With 37%, Jerry, I tell you, with 37% probability, it's bound to get me the best of the potential brides. It's mathematics, Jerry, something called the Secretary Problem. And let me tell you, if one or two happen to be a secretary, I wouldn't mind at all. Anyways, that's what I'm going to do.

Jerry: Good plan, but there's one problem.

George: What's that?

Jerry: What do you think the chances are that you'll find 20 women who would be willing to date you?

CUT!

RATIONALITY

How do people ultimately make a decision in the Secretary Problem? What happens most often is that they don't aim to pick the best candidate, but rather the first candidate who is above their threshold for acceptability. This raises the issue of how people actually make decisions, as opposed to how they *ought* to make decisions. There are those who argue that our earlier proposal of maximizing expected utility may not make sense in decisions that are "one-offs," single decisions that are never repeated, because expectations are only average values over many plays of the same game. There are also those who feel that we humans lack the ability and desire to find optimal solutions, but rather settle for the best solution that meets some minimum needs (such decision making is called *satisficing*, a mix of "satisfying" and "sufficing"). There is a strong case to be made for using satisficing in many decisions we make – finding all the available options may require a huge effort. We often say to ourselves, "If I find something that meets my needs or goals, I'll settle for that."

However, there are times when we seek the very best option, and maximizing expected utility has found many proponents. What is important, I think, is that you have a process for making your decision, one that has some reasoning behind it. How much do you value what you value? How likely is each possible outcome? Are you a *maximizer*, wanting to get the most out of each decision, or a *satisficer*, willing to accept less as long as the outcome is acceptable? I think there is a lot to be gained by carefully considering your approach to decision making, whether you are making big life decisions like changing careers or moving cities, or smaller ones, like selling your house. How we make decisions is part of how we view ourselves. Wouldn't you like to view yourself as a rational decision maker rather than as a by-the-seat-of-your-pants one?

IS THAT ALL THERE IS?

Life would be so simple if we needed to consider only our own choices and decisions. But it gets trickier when others are involved, each making their own decisions. This takes us into the realm of what mathematicians call *game theory*, where two or more players compete, each with their own choices, outcomes, and values. The theory of games originated in the mid-1940s, in the joint work of an economist, Oskar Morgenstern, and a brilliant mathematician, John von Neumann (the latter also worked on the development of the atomic bomb).

The games we play range from the simple (tic-tac-toe and rock-paper-scissors) to the complex (chess). But we can also view our other competitions with one another as games.

ROCK-PAPER-SCISSORS

Many games are two-person games, and if they are completely competitive (what I lose you gain and vice versa) they are called *zero-sum*. If you consider wins positive and losses negative, then the sum of what I get and you get must always be zero. Most games we play are too complex to unravel by using mathematics. One that isn't is the age-old game of rock-paper-scissors. Remember? Rock breaks scissors, scissors cut paper, paper covers rock; if you and your opponent show the same hand sign, it's a draw. For those with wide-open social calendars, there are even rock-paper-scissors tournaments.

Is there a best way to play? Well, it's clear that there is no single *pure* strategy (a strategy to play just one of the options) that's going to work over and over again with an opponent. For example, if you always play rock, your opponent will figure it out after a few games and always play paper to win.

The best way to play is to mix up your pure strategies and play rock one-third of the time, scissors another third, and paper for the rest. Then, no matter what your opponent does, you will win on average at least one-third of the time. Mathematics shows that this is the best result you can hope for. What fascinates me is that a game with clear rules can have an optimal strategy that still involves randomness in play.

THE NEED FOR COOPERATION

I see games where most people wouldn't. As I'm filling up my minivan with gas, I look at the price per litre and think of how it's set in a free market. All of the surrounding gas stations need to fix a price for their gasoline. In the absence of any collusion, how should it be set? If I were a station owner, I'd look around and set my gas price a little lower than everyone else's. That way I'd "win" the game and more customers would flock to my gas station.

But the other station owners are probably as rational as I am and would do the same thing; a price war would inevitably result. Where would the price settle? Mathematically, the price should dwindle right down to a penny per litre above the cost to retailers. It is for this reason that the owners of any commodity consciously or unconsciously collude to "fix" the price at a reasonable level, so all can maintain some reasonable profit margin. Of course, a gas station owner who is new to the game, or who is unwilling to accept the status quo, can start a price war, but soon everyone is a loser – except the public!

COUPLEDOM NEED NOT BE A BATTLE

Ah, the Battle of the Sexes. My wife and I have been trying to decide where to go on vacation. She would like to go to Europe and visit historic landmarks. I would rather travel to Cleveland, relax, take in some ballgames, and go to the Rock and Roll Hall of Fame. What to do? I'd like to come up with a rational decision we can agree on. My wife often claims that in our arguments I am too "professorial," but I think she is just sore because when we argue, I sometimes don't call on her when she raises her hand to speak . . .

In the business deal versus romantic dinner example, I drew a tree to illustrate the one-person decision-making process. For our two-person holiday conundrum, we can form a table for the decision process as follows (note that the utilities I assign here, and elsewhere, are subjective, but they probably aren't too far off how you and yours would assign them):

	MY WIFE'S CHOICE	
	Go to Cleveland	Go to Europe
MY CHOICE		
Go to Cleveland	My utility: 100 Wife's utility: 10	My utility: –20 Wife's utility: –20
Go to Europe	My utility: –100 Wife's utility: –100	My utility: 10 Wife's utility: 100

What the table indicates is the utility (i.e., value) of each of the outcomes for myself and my spouse. So, for example, if my choice is to go to Cleveland and my wife's choice is to go to Europe, we look at the first row, labelled "Go to Cleveland" (as that is my choice), and at the second column, labelled "Go to Europe" (as that is my better half's choice). This takes us to the box in the first row, second column, which states "My utility: –20, Wife's utility: –20"; this means that the combined choice of my going to Cleveland and my wife going to Europe gives us each a utility of –20, not a great choice for either of us.

Now here is the problem. I want my wife to join me in Cleveland, and my wife likewise wants me to go with her to Europe. Neither of us really wants to be alone for the vacation, as indicated by the negative utilities for the two cases where we go our separate ways. (I even allow for the outcome that we both end up at the other's vacation spot, perhaps deciding at the last moment to switch our separate travel plans from our favourite destination to please the other.) The actual utilities are critical, but their sizes relative to one another are even more important.

Now we see the problem. Both choices to go together to either vacation spot (the upper left and lower right boxes in the table) are rather stable, in that we wouldn't decide on our own to change our plans, as it would only decrease our utility. Likewise, the two choices where we end up in different locations are unstable, as one of us can increase our utility by switching choices. What we end up with is a dilemma – travelling together makes sense, but which trip to choose? In this type of two-person game, one person could try to threaten the other (not with violence, of course). If I were a cad, I could say to my beloved: "Look, if you insist on us going to Europe, I'm going to change my mind and go to Cleveland; even though it will be worse for me, it will be much worse for you. So you should agree to go to Cleveland with me to avoid all that." But both sides can make threats, which can lead to negative outcomes. There seems to be no rational way to make a choice, except perhaps to agree to alternate vacation spots over the years.

But whether we travel to Europe or Cleveland, we'll both get something out of the trip, as both our utilities are positive in both cases. But we'll see in a moment that there are other situations where we can paradoxically be driven to the worst mutual outcome, no matter how rational we intend to be.

CHICKEN? BOCK, BOCK!

Competition often brings to mind another game from my childhood. Remember the game of chicken? One of my siblings and I would start to run toward each other and the one who veered out of the way first was deemed the "chicken." Of course, if neither veered out of the way, we'd collide with a painful thump and we would declare a tie (for some reason I remember very few ties!). Nowadays, I still sometimes seem to get caught in a game of chicken. When in the midst of an argument, do I back off or proceed full steam ahead? I remember on a recent trip I saw an example of road rage close up. Two drivers, one in his twenties and the other in his forties, jumped out of their cars and stood face to face, screaming and staring fiercely at one another. Here is what the utilities might look like for each combatant:

	OLDER GUY'S CHOICE	
	Confront	*Back off*
YOUNGER GUY'S CHOICE		
Confront	OG's utility: –200 YG's utility: –200	OG's utility: –100 YG's utility: 100
Back off	OG's utility: 100 YG's utility: –100	OG's utility: 50 YG's utility: 50

Note that neither man is eager for a knock-down fight, as they don't know how crazy the other is and how badly they might get hurt. On the other hand, there is a definite advantage to a confrontation if the other person backs down: the aggressor feels great and is a hero to onlookers, while the person who backs down has to drop his tail and shuffle off. The real issue is that the best joint decision is to de-escalate the situation and back down, but the solution is *unstable* in that there's a very strong inclination to escalate if you think the other person will back down.

I resisted the temptation to jump out and explain the theory of the game of chicken to the drivers, and left them to sort it out. Many escalations in life can be viewed as a game of chicken, and repeated exposure to the game ought to lead to a recognition of the need for co-operation.

THE GLOBAL WARMING GAME

Global warming is in the news practically every day. Even though the dire consequences seem decades away, a mathematical perspective condenses time and makes the future seem as immediate as the present.

I see global warming as a game played among countries. Each country has the choice of whether to reduce greenhouse gas emissions or not. There is a cost to such a reduction, but a mutual benefit as well. To simplify, let's consider the situation with only two countries, Country A and Country B. The issues that become evident apply to a multiple-country situation as well.

Should governments legislate corporations and individuals to reduce emissions or not? It's not as straightforward a choice as it seems. I've drawn another table and attached what I see as some reasonable utilities for each country. (The precise utilities would, in practice, be unknown, but would likely follow a similar pattern: the utility for a country would increase significantly if it didn't reduce its emissions, while the other country did, and the utility would be lowest when both countries failed to reduce emissions):

	COUNTRY B'S CHOICE	
	Reduce emissions	*Don't reduce emissions*
COUNTRY A'S CHOICE		
Reduce emissions	A's utility: 1,000 B's utility: 1,000	A's utility: −10,000 B's utility: 10,000
Don't reduce emissions	A's utility: 10,000 B's utility: −10,000	A's utility: −5,000 B's utility: −5,000

It would be good for all citizens and governments if both countries reduced emissions, as indicated by the utilities of 1,000 in the table. But here is the problem: if both countries realize this, then there is an incentive for one country to change its mind and not reduce emissions, as it will benefit economically by not having to invest in reducing greenhouse gas emissions (and hence has a higher utility, which I pegged at 10,000). The country that remains determined to reduce emissions will actually have its utility reduced by doing so, as it will pay higher costs while possibly reaping fewer benefits while the other country continues to emit. So what happens? Both countries continue to emit greenhouse gases, a choice that hurts everyone (the utility of doing so is far lower than if they both reduce emissions). It makes no difference how much the countries negotiate; as long as each acts in its own self-interest, the outcome will be the same. Although there may be long-term contracts and agreements to reduce emissions, if they are not enforceable, everyone will cheat.

Furthermore, even yearly agreements fail to solve the problem. Should a country decide to abide by the agreement? You would think this is a simple decision, but consider the following mathematical argument that each country should *always* cheat and break the agreement if it thinks the planet is doomed. If they are right, there will be one last agreement to sign, which, precisely because it is the last agreement and hence unenforceable as earth is doomed from that point on, both countries will break the agreement. But this essentially makes the second-last agreement the last one, and we know what happens with the last agreement: both parties will break it. Again, both parties will break the second-last agreement for the same reason, and we proceed back in time until we realize that every agreement, from the first one on, will be broken.

This global-warming "game" is an example of the *Prisoner's Dilemma*, an old paradox dating back to 1950. Two mathematicians, Merrill Flood and Melvin Dresher, discovered the problem while working at the RAND Corporation, a U.S. think tank, and another mathematician, Albert Tucker, soon rephrased it and gave it its famous name. In the scenario imagined by Flood, Dresher, and Tucker, two prisoners are each offered a choice: confess or profess innocence to a major crime that the prosecutor thinks they have both committed. If they both confess, they will each end up with a prison term of 10 years. If neither confesses, they will each be convicted of a lesser crime and sentenced to one year. But if one prisoner confesses and the other doesn't, the one who has confessed is viewed as turning state's evidence and is released while the book is thrown at the other – he will get 20 years in prison. Think the problem through; no matter what discussions transpire between the prisoners, inevitably both break rank and confess at the last moment. The Prisoner's Dilemma is recognized for its inherent paradox, that two rational beings inevitably end up with what is the worst possible choice.

So is there a way out of our global-warming dilemma, or are we doomed? Is there any room for altruism? Well, there are a number of tacks we can take. One way that has worked very well in tournaments of the Prisoner's Dilemma (yes, there are such things!) is the "tit for tat" strategy, where you reply in kind to your opponent's move in the last instance of the game. If she has co-operated, you co-operate, and if she has betrayed you, you betray her. This strategy was first proposed by a game theorist, Anatol Rapoport, and seems to work quite well when games of Prisoner's Dilemma are played successively. It does require a good dose of forgiveness on the part of the participants, and is not useful when only a single instance of the game is played.

Another way to deal your way out of the Prisoner's Dilemma is to change the utilities. If we let our governments know that we highly value reducing greenhouse gas emissions, no matter what, then the utility to continue to emit will drop, even in the face of other countries that continue to pollute. If governments sense the utilities have changed, as in the following table, then what results is a stable solution, namely for both countries to reduce emissions; neither country has any incentive to switch to another choice. Then we can all join together and declare that defeating global warming is not only the right decision, but the rational one as well.

COUNTRY B'S CHOICE

	Reduce emissions	Don't reduce emissions
COUNTRY A'S CHOICE		
Reduce emissions	A's utility: 1,000 B's utility: 1,000	A's utility: −100 B's utility: −100
Don't reduce emissions	A's utility: −100 B's utility: −100	A's utility: −5,000 B's utility: −5,000

Not every problem is as weighty as global warming, nor, thankfully, as difficult to solve. Often it's too easy to make a rash decision, in the hope that a quick decision will turn out to be a good one. But life is a series of choices – whom to marry, whether to worry, how to do business. There are some forks in the road where the decision you make can influence the rest of your life. You need not be alone at such critical points: mathematics can be a trusted advisor along the way.

My wife's plane finally lands safely, just as the odds say it should. After a hug and a kiss and a "thank-goodness-you're-here-now-you-can-take-the-kids-please!" we're on our way back home. This time I make sure that there is no more math research dancing a cha-cha in my head, and that I drive at the speed limit. That's one decision that's easy to make.

REDECORATING AND GEOMETRY

On the drive back from the airport, my wife is talking about renovating the house. We are having a big family get-together this summer in Halifax, and she wants to finish the basement and repaint the main floor. Like most pure mathematicians, I'm not adept at such applied tasks, due to lack of practice. Given the choice, I choose cerebral work over the physical kind every time.

Still, planning is fun, and working with layouts is just like plane geometry. Doing quick sketches on napkins is reminiscent of the old days of mathematics. (I used to actually do my research on napkins in between sets while playing guitar in bars. Some of my best results were lost to a sneeze.) I like that I can check some of the estimates for materials, to keep the contractors honest. I don't want to be caught in a reno-gone-wrong story that lasts for years!

I'm a very visual guy. I like to see what I am doing. I used to enjoy art a lot in school, especially sketching. This tendency even influenced what area of mathematics I chose to do research in; networks are, for most purposes, visual beasts. Pictures are an essential part of mathematics, and trying to do mathematics without them is throwing away all of the understanding and intuition.

I remember plane geometry from elementary school – triangles moving around on the overhead projector. It was probably my first introduction to proofs, though I have to admit that geometry never really did catch my fancy until I hit trigonometry. The ability to calculate angles from distances and vice versa seemed pretty cool to me.

Geometry is one of the oldest branches of mathematics, and certainly the oldest one that stresses deductive proofs. Geometry dates back to two ancient Greek mathematicians, Pythagoras and Euclid. The latter put plane geometry (that is, geometry on a flat surface) on a firm mathematical basis. He postulated a system of five axioms, or assumptions, things like "You can draw a line through any two points." These axioms may seem obvious . . . but more about that later. Euclidean geometry allows us to prove basic geometrical facts – that two particular triangles are *congruent*, meaning that they have exactly the same shape, and that they are *similar*, meaning that they have the same angles and are scaled images of one another.

THE CORNER POCKET

When I was growing up we had a pool table in the basement, and I spent many an hour playing with my younger siblings. I have found that people with a mathematical inclination often enjoy pool. Wolfgang Amadeus Mozart loved both mathematics and billiards.

Early on I learned a law of physics that stated that an object that strikes a flat surface leaves at the same angle it arrives at. This *Law of Reflection* had obvious implications for shooting pool, in that the angle the pool ball leaves a cushion is the same as the angle it strikes it at (provided you don't hit the cue ball with any spin, or "english," as it's known). I often used to carry a protractor from my geometry set with me and measure various angles on the table. I did this for two reasons. First of all, it helped my game. I could plan to get out of "hooks" by a number of bank shots. And more importantly, it drove my sister crazy. Ah, the joys of childhood.

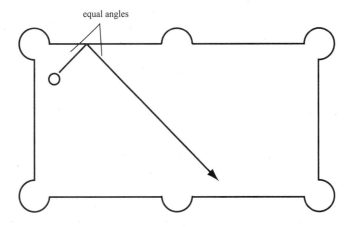

equal angles

My use of a protractor was very rudimentary, and there are devices known as *inclinometers* or *tiltmeters* that are used by surveyors, engineers, and others to measure angles. One of my favourite uses of angles involves measuring large heights.

Suppose a surveyor with an inclinometer and tape measure wants to measure the height of a building, but can't do it directly. She can first measure accurately the distance from where she is standing to the base of the building; let's suppose it's 15 metres. Now the surveyor sets up an inclinometer to measure the angle up to the top of the building, from her eye level, say 1.8 metres. If the angle measures 58.6 degrees, then a quick sketch looks like this.

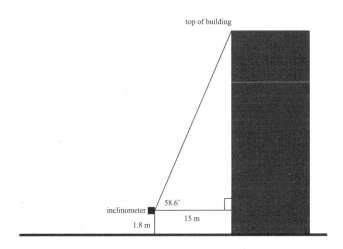

What becomes clear from the picture is that we have a right-angled triangle. Now here is where those hours of trigonometry come into play. Remember learning about right-angled triangles and sines, cosines, and tangents? What we want to determine is the length of the side opposite the known angle, 58.6°, and we know the length of the side *adjacent* to the angle, namely 15 metres. Trigonometry tells us that the tangent of the angle is equal to the opposite over the adjacent: $\tan(58.6°) = \dfrac{\text{opposite}}{\text{adjacent}} = \dfrac{\text{opposite}}{15 \text{ metres}}$. A calculator will tell you (if you have it set to degree mode) that the tangent of 58.6

degrees is about 1.638, so by cross-multiplying the 15 metres, we get 1.638×15 metres=24.57 metres. That is the height of the building from the level of the inclinometer to the top of the building. Since the inclinometer is 1.8 metres off the ground, the total height of the building is 24.57 metres+1.8 metres=26.37 metres.

If you don't have an inclinometer available, you can use a protractor, with the centre right beside your eye and the bottom level to the ground, though the result will not be as accurate. But calculating heights and lengths via trigonometry makes for a fun activity with your kids (really!), and there is some neat mathematics to learn (and recall) along the way.

MATH RENOVATIONS

Math comes in handy when estimating costs for home renovations. My wife and I are thinking about painting the dining room. I took some measurements and found that the room has a height of 95 inches. The room has the shape of a pentagon, with an open entranceway and window, as shown:

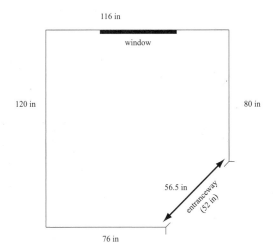

The large window together with its frame (which we won't paint now) is 60 inches by 65 inches. The entranceway together with its frame is 52 inches by 83 inches.

I fill in a couple of measurements by extending the shape of the room to a rectangle. This extension of a geometric figure using imaginary lines is a common trick in geometry.

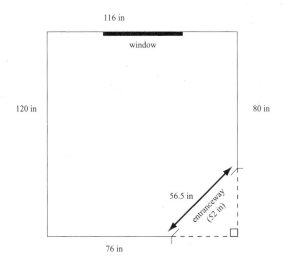

I see now that the 56.5-inch wall, with the entranceway, hits its adjacent walls at the same angle, as the two missing sides of the right-angled triangle have the same length. Since the inside angles of the triangle are equal, the angles on the other side of the triangle will be equal to each other as well. In fact, the angles where the walls meet are both 135°. Why? Because the two acute angles in the right-angled triangle add up to 90° (the three angles in a right-angled triangle always add up to 180°), and as they are equal to each other, they both must be 45°. This means that the angles on the other side of the lines are 180° − 45° = 135°.

Now the area of the walls is the perimeter of the room times the height. The perimeter of the room is 56.5+76+120+116+80=448.5 inches, and the height is 95 inches, so the wall area is 448.5 in × 95 in = 42,607.5 in^2. This includes the entrance-way and the window, however, which have areas of 52 in × 83 in=4,316 in^2 and 60 in×65 in=3,900 in^2, respectively. So the total area we need to paint is 42,607.5 in^2 − 4,316 in^2 − 3,900 in^2 = 34,391.5 in^2. But in order to buy the right amount of paint, we need everything to be in square feet, so we use a conversion factor, as I talked about in Chapter 2:

$$34,391.5 \text{ in}^2 \times \frac{1 \text{ ft}}{12 \text{ in}} \times \frac{1 \text{ ft}}{12 \text{ in}} = 238.8 \text{ ft}^2.$$

We get about 250 square feet of coverage from a gallon of paint. We can think of this as stating a conversion factor from square feet to gallons of paint. We find that we need

$$238.8 \text{ ft}^2 \times \frac{1 \text{ gallon}}{250 \text{ ft}^2} = 0.96 \text{ gallons of paint.}$$

Of course, we can't buy this amount, but we can buy a gallon, only a little bit more than we need.

If we decide to paint the ceiling, then the dotted lines in the diagram come in really handy. The second diagram shows that we can view the ceiling as a rectangle minus a right-angled triangle, instead of a pentagon, and that simplifies things greatly. The big rectangle has an area of 116 in×120 in=13,920 in^2. We have over-counted by the area of the little triangle, which has a base and height of 40 inches. The area of a triangle is half the base times the height, or, in this case, $\frac{1}{2} \times 40$ in × 40 in = 800 in^2.

Therefore, the area of the ceiling is the difference of the two numbers, that is, $13,920 \text{ in}^2 - 800 \text{ in}^2 = 13,120 \text{ in}^2$, which we again can convert to 91.11 square feet. This corresponds to

$$91.11 \text{ ft}^2 \times \frac{1 \text{ gallon}}{250 \text{ ft}^2} = 0.36 \text{ gallons},$$

so we'll need a couple of quarts of whatever paint we're going to use when we do the ceiling (there are four quarts to a gallon).

A SHORT BREAK FOR SOME CAKE!

A big storm crept up the coast from the eastern seaboard last week and covered Halifax with a sheet of ice. Both of my sons were home, with school being cancelled for the day, yet again. They were hungry and wanted some cake. There was, of course, just one piece left. If you have children, you know that you must be seen to divide the cake *exactly* evenly. Mathematics tells you that you can always cut the cake equally, no matter what shape the cake is. (One year my wife and I tried to cut a cake and reshape it into a *Tyrannosaurus rex*, the birthday boy's favourite dinosaur. It was absolutely horrible: none of the realigned pieces stuck together; the cake completely fell apart. I had the idea to take some red icing and convert the cake's remains into the T. Rex's first kill. My son loved it, and victory was snatched from the hands of defeat.)

The argument that you can always cut a cake evenly is pretty easy to prove. Start with the knife out to the left of the cake, at any angle that you like, so that all of the cake is to the right. If you move the knife to the right over the cake, the amount of cake to the right of the knife decreases. The percentage of the cake that is

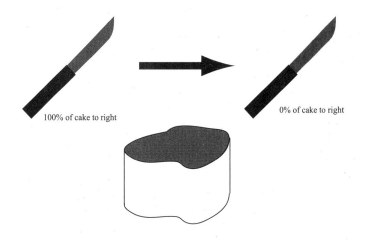

to the right of the knife changes continuously, from 100% to 0%, and at some point along the way there is exactly 50% of the cake to the right. That is where you take your slice if you have two equally hungry kids. If you have more kids, the same idea applies but with a different percentage, and more cuts.

100% of cake to right

0% of cake to right

This notion of continuity is a very powerful one. For instance, imagine yourself climbing up a mountain from noon to midnight, then turning around and walking down from midnight to noon. There would be a time, in hours, minutes, and seconds (ignoring A.M. and P.M.) at which you were at the same altitude on the trip up as on the trip down. For if you plotted your ascent over the 12 hours, you would get a nice continuous curve from top to bottom. Plotting your descent is equally continuous, but runs from the top height to the bottom. You can see from the picture that the two curves have to cross, and at that point you'll be at the same altitude and at the same time.

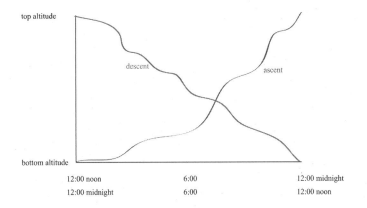

top altitude

descent

ascent

bottom altitude

| 12:00 noon | 6:00 | 12:00 midnight |
| 12:00 midnight | 6:00 | 12:00 noon |

But back to the cake. I could have cut the cake into two equal pieces. On the other hand, the boys were in the other room, and that cake did look tasty . . .

IN THE CAN

I was going to leave the cake for the boys, really. But I did want to nibble on something. I am trying to lose a bit of weight, so junk food is out of the question. In the pantry there are a large number of cans. What to choose?

A question occurs to me: What determines, or should determine, the shape of a can of food? This is a math problem I can sink my teeth into, and it takes my mind off eating for the time being.

All of the cans are *cylindrical*, having identical metal circles on the top and on the bottom, and a metal cylinder joining them. I am assuming that the manufacturer would have a certain volume of food that he would like to fill cans with and that he would want to minimize the amount of metal used in each can, thereby lowering his costs. I am also assuming that the cost is directly proportional to the amount of metal used, and that the top and bottom are made up of the same material as the sides. I am also ignoring the little rims that run along the top and bottom of the cans. Whenever I try to solve a problem with mathematics, I always need to make some assumptions, which I hope are reasonable. There is a principle, which most mathematicians ascribe to, called *Occam's Razor* that states that when you create a model or a theory, simplest is best.

Back to the problem. The volume of the can is the area of the top circle times the height. I draw a picture and label the radius of the top circle r, the height of the can h; I call the volume V and the surface area S. I see that what I want to minimize is the surface area, which is the combined area of the top and bottom circles plus the area of the side. The latter looks tricky, but if I imagine slicing the can open vertically, it unrolls into a flat sheet, with one side being the height of the can and the other being the edge around the bottom (or top) circle, that is, the *circumference* of the circles, which I remember from school to be $2\pi r$ (the Greek symbol *pi*, π, is about 3.1416).

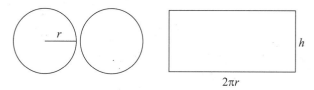

The areas of each of the two circles are easy, πr^2. The area of the side of the can is now also easy – it is a rectangle of dimensions $2\pi r$ by h. So the surface area I want to make as small as possible is $S=\pi r^2+\pi r^2+(2\pi r \times h)=2\pi r^2+2\pi rh$. But the constraint I have is that the volume is fixed. The volume of a cylindrical can is the area of the bottom times the height. The bottom is a circle, whose area is πr^2, and the height is h. So we have $V=\pi r^2h$. V is what is known as a *parameter* or *constant*; it changes in different instances of the problem, but in each instance it is some fixed number. If we solve this equation for h, which is one of the things we *don't* know (the other is the radius r), and substitute back into the formula for the surface area of the can to get rid of one variable, h, I find that what I want to do is minimize $S=2\pi r^2+\dfrac{2V}{r}$ where r can be any positive number (if r is small, the can will be tall and thin, and if r is large, the can will be short and wide).

This is exactly the type of problem that calculus is built for, and with a few steps I find that the optimal shape for the can to have is $h=2r$, that is, for the height of the can to be equal to the diameter of the bottom circle (the diameter of a circle is twice the radius). The can that I did pull from the cupboard was full of pineapple, and a quick measurement shows that indeed the diameter and the height of the can are almost the same.

At the supermarket I can see that not all cans have the optimum shape, but other considerations may come into play. Taller cans are easier to grab, for example. The solution also didn't take into account different costs of materials for the top, bottom, and sides of the can, the lips around them, or the cost of the label. Well-known statistician George Box is often quoted as having said, "All models are wrong, but some are useful."

Still, my calculation gives me a pretty good idea of what a can's dimensions should be, and it was fun. As I finish the last of the pineapple, I wonder: was I hungry for pineapple, or did I choose the can because it had the right dimensions? Food for thought.

WHERE IN THE WORLD IS JASON BROWN?

Geometry took a quantum step forward when the mathematician and philosopher René Descartes invented what is known as the *Cartesian coordinate system*. This system labels points by a pair of coordinates (for flat, two-dimensional geometry) and a triple of numbers (for three dimensions), all measured from a fixed central point, called the *origin*, which is labelled with zeros: (0,0) or (0,0,0). Where the origin is located is problem-dependent, often using the most convenient natural choice of where to start your measurements. Cartesian coordinates are taught early in school, and it is hard to imagine that at one time such a concept didn't exist. Much of mathematics is like that; ideas that were once unknown were discovered or invented by brilliant mathematicians, only to be filtered down over the centuries to school-age children as "obvious" mathematics.

Maps are essentially Cartesian coordinates, but with one side labelled with letters rather than numbers. I can remember from a very young age sitting in the back of my parents' car and following along on a map. I still have a pretty good sense of direction; I can almost always point upwards and downwards when necessary. When my wife and I used to travel, she would do most of the driving and I would navigate. We still do that, but now many rental cars have a Global Positioning System (GPS), which helps to some extent. I still miss having a map in my hands, though.

I recently came back from a trip to Cambridge, England, where I attended a research workshop and gave a lecture at the Isaac Newton Institute for Mathematical Sciences. This was my first trip on which I relied completely on a GPS system, and it made driving on the left side of the road almost bearable. I don't know how the English do it. I thought that holding up a mirror would help, but sadly no.

Anyway, the GPS was so reliable that I wanted to find out more about how it works. There are many satellites spinning around the earth, specifically for the purpose of GPS. They each send out a signal that the GPS device in your car receives. The signals are stamped with the time they were sent and the location of the satellite, and the GPS receiver uses its internal clock to calculate the time it took for the signal to reach your location. Having calculated the time, your receiver multiplies it by the speed of transmission to get the distance of the satellite from the receiver.

Any GPS system manual will tell you that you need to have a direct line of sight to at least four of the satellites orbiting earth (which is why GPS systems can't be trusted in tunnels). Knowing only your distance from a fixed location (a satellite's position in space when it sent the message) is not enough to tell you where you are.

In order to understand why, suppose that I am hiding and all you can determine is my distance from a fixed location. If you know that I am 1.8 kilometres away from the lamppost in the centre of town, all this would tell you is that I am on the boundary of a circle of radius 1.8 kilometres from the lamppost; I could be anywhere on the circle.

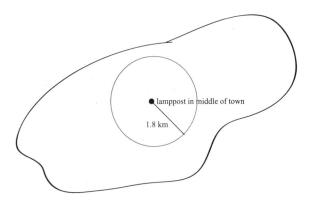

Now suppose you also know the distance I was from a second spot in town, say another lamppost on the town's limits. If you are told that this new distance is 3.4 kilometres, then I would have to be on the boundary of the second circle, centred at the lamppost at the edge of town, with radius 3.4 kilometres. This would narrow things down quite a bit, but there would still likely be two possible locations where I might be – either of the two places where the two circles meet indicated by question marks in the diagram.

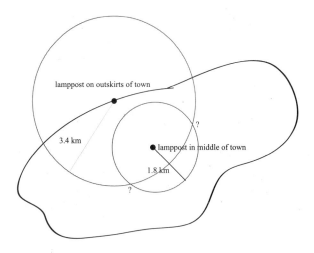

You'll need a final, third location from which to measure my distance in order to fix my location: let's say 2.8 km from a third lamppost somewhere else in town. There will only be one point that is in common among the three circles, and I must be there. You found me!

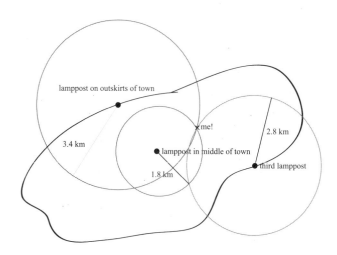

Finding my location in three dimensions is much the same, except that the set of points at some fixed distance from a central location is a sphere rather than a circle. In general, the intersection of two spheres is a circle, the intersection of three spheres is two points, and the intersection of four spheres, provided they intersect, is a single point. This is why a GPS system needs to be able to gather data from at least four satellites to pinpoint where you are; with fewer satellites, it has to make a guess. All of this tells me that GPS should work whether I'm on the surface of the earth or in an aircraft; the intersection of four spheres locates objects anywhere in three dimensions. That's good to know in case I'm lost in space (which my wife often claims I am).

LOCATION, LOCATION, LOCATION

The process of locating a position by using the distances from fixed points is called *triangulation*, and at one time, the easiest way to do it was with a compass (the kind from a geometry set) and a map. You put the pointer at one of the fixed points, measure out the distance from that point, and draw a circle. When you repeat this action from two other points, the intersection of the three circles gives the position of the object.

Giving directions using a directional compass is another model of navigation, different from Cartesian coordinates. Measuring the angle and distance from a fixed point is the basis of *polar coordinates* in mathematics. Mathematicians tend to measure angles counterclockwise rather than clockwise, as normal people do, but the idea is the same. There are ways to switch back and forth between polar and Cartesian coordinates using trigonometry, but which coordinates you use depends on the ease of measurement and what you're using the measurement for.

Polar coordinates are very useful when what is important are the distance and the angle. I just bought a new microphone for recording some music, and the specification for the response of the microphone to sound is shown in polar coordinates. In fact, the microphone is called a "cardioid condenser microphone," as the polar pattern for pickup of sound is a cardioid curve in polar coordinates (I kid you not; this is on the specifications that are included on the box). A cardioid is a curve in polar coordinates that is heart-shaped (hence the *cardio* part of the name). The equation of a cardioid is pretty straightforward in polar coordinates (for example, $r=2+2\sin\theta$), and would be much more complicated to write down in Cartesian coordinates. Circles can be described with an equation of the form r equals a number, and lines through the centre are just the angle (usually written with the Greek symbol θ) equal to a number. What I can read from the diagram is that the microphone picks up sounds furthest from the source when the source is directly in front of the microphone (the top of the picture) and that sounds to the sides and behind the microphone are not picked up as well.

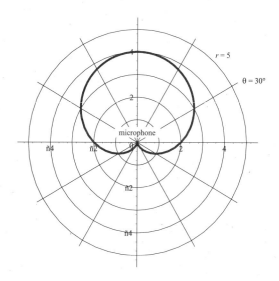

Polar coordinates are also good for giving directions. If you are trying to direct someone to a house that is four kilometres east and four kilometres north of where you are, it's probably easier to tell a person with a compass to go 5.66 kilometres northeast. (Though for someone in a car, going east, then north may be preferable. Different horses for different courses, as mathematician Paul Erdös used to say, referring to the fact that when choosing a colleague to do research with, he always selected a problem suitable for that person's interests and skills.)

In addition to wanting to locate objects, we often want to talk about motion and *vectors*. A vector is a mathematical object that has a direction and a length.

This model is particularly well suited to air travel, as the effect of the wind is easily taken into account using vectors. If I am lucky enough to get a flight where I have a seatback screen, I follow the tracking of the aircraft, which is shown as a vector on the screen. One of the beauties of mathematics is that what we call a "length" can be anything, and in navigation, the length is often the speed of the object. So if an airplane is travelling northeast at 880 kilometres per hour, the motion could be represented by a vector that has length 880 in the direction northeast. Pictorially, a vector is usually drawn as an arrow in the Cartesian coordinate system.

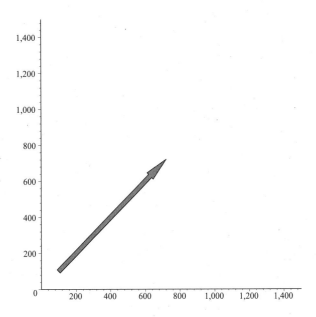

There is another way to keep track of two-dimensional vectors, and that is by writing down a pair of numbers – how much you are going to the right, and how much upwards. By measuring (or using a little trigonometry), you can find that the vector for the plane was drawn starting at the point (100,100) and ending at (722,722). The vector is the difference between the pairs: (722,722)–(100,100)=(622,622).

The real advantage of vectors is how they can be added together. If there is, say, a headwind from the north at 58 knots, or 107 kilometres per hour, then this will blow the plane off track, in a southerly direction. The question is by how much.

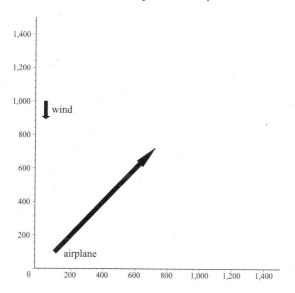

You need to know that you add vectors "tail-to-head," that is, you take one vector and attach its tail to the tip of the arrowhead of the other. The resultant vector travels from the tail of the first to the head of the second.

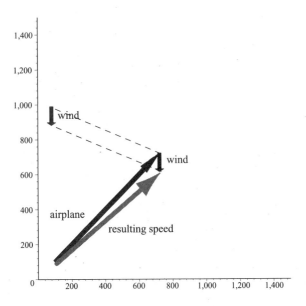

In order to figure out exactly how much the wind affects the speed of the aircraft, you could draw it out on graph paper and measure the direction and length of the resulting arrow. A far easier way is to keep track, for every vector, of how much you go to the right (east) and up (north), writing the answers as pairs. As we established earlier, the plane's vector is (622,622), that is, the aircraft is going 622 km/hr east and 622 km/hr north, and the vector for the wind is (0,−107) (down is negative/south, as is left/west). Now all you do is add the vectors in the obvious way: (622,622)+(0,−107)=(622+0,622−107)=(622,515), which tells you that the plane will end up going 622 km/hr eastward and 515 km/hr northward, that is, the resulting vector is (622,515). Of course, you'd like an overall speed and direction. You get the length of a vector from that old chestnut, the Pythagorean theorem; the length is the square root of the sum of the squares of the components of the vector. In our case, it is $\sqrt{622^2+515^2}=808$ km/hr. The direction is again a bit of trig, and is in fact about 40 degrees up from east, which of course is five degrees south of northeast.

Pilots use vectors not to calculate what path they are actually travelling, given the wind speed and direction, but to alter their path so that they head in the right direction, taking into account the wind. But the idea is the same; vectors, those hybrids of geometry and algebra, rule.

My wife and I have decided to hire contractors to finish the basement – I think I had best leave that to the experts. I can lend a hand when it comes to planning, though. Mathematics helps to ensure that we don't do things by the seat of our pants, except when calculating. Measure twice, renovate once. At least I hope that's how it works!

8

A LONG TAIL, OR THE ONE THAT DIDN'T GET AWAY!

After dropping my wife at home, I head back into work at the university. I've spent the last hour or so catching up on my administrative work.

I've got to take a break from all the paper shuffling. I scan an online magazine, where there is an article about the "long tail." No, it's not what those in charge of setting credit-card rates possess. It has to do with distributions and sums of many, many small numbers. I put down my coffee mug; someone at the magazine knows a bit of mathematics.

The discussion of the long tail brings back memories of infinite series, one of my early loves in mathematics. Making sense out of adding up infinitely many tiny numbers captured the attention of many a mathematician over several centuries. Pretty, pretty, pretty. And pretty useful, too.

The sum of small bits must be small, right? Well, not always. The result of adding up many small bits can turn out to be quite substantial. I remember watching one of the *Superman* movies, the one with Richard Pryor in it. As I recall, at the beginning of the movie, Richard Pryor is caught stealing millions of dollars from the U.S. government, by siphoning off all of the fractions of cents that are rounded off of government paycheques. The many millions of fractions of cents added up to a whopping sum.

This comes to mind today as I read about one of the growing effects of the Internet called "the long tail." The idea is as follows. You have some distribution of numbers. It might represent demand for what your company produces (say, CDs by certain bands) in various markets. If you are lucky and mainstream, then in the major markets you might have high demand, and it is worthwhile to produce a lot of CDs. The sales will be there. On the other hand, if what you manufacture has only a niche market, with relatively small demand, it probably isn't worthwhile to produce at all. Costs for distribution to outlying markets eat up too much of the profit.

But imagine the tail of the distribution is long – that is, there may be only a few people in each market who want what you produce, but there are a lot of markets across the world. The sums of many small amounts is not insignificant, and the growth of the Internet has made this "long tail" accessible to producers. It is now viable for small companies to produce items (perhaps even minimizing distribution costs if they can be distributed over the World Wide Web) for the sum of the markets. The math is there.

SUMMING TO INFINITY

This notion of adding up many, many small numbers goes back once again to the ancient Greeks. One of the most famous paradoxes was formulated by Xeno, and has to do with a race between Achilles (he of the bad heel) and a tortoise.

Anyway, as the paradox goes, Achilles runs at 10 kilometres per hour, while the tortoise runs much slower, one kilometre per hour. Achilles, the nice guy that he is, gives the tortoise a head start of 10 kilometres. We know that Achilles will eventually pass the tortoise, as he is running faster. In fact, after two hours Achilles has run 20 kilometres, while the tortoise has run only two kilometres, so that Achilles is at the 20-kilometre mark, while the tortoise is only at the 12-kilometre mark. But consider the point when Achilles makes it to where the tortoise started; this happens at exactly one hour into the race, as the tortoise was ahead by 10 kilometres to begin with, and Achilles runs at 10 kilometres per hour. The tortoise is then ahead by the one kilometre he has run since the start.

It takes Achilles one-tenth of an hour to reach the 11-kilometre mark (the time taken is the distance, one kilometre, divided by the speed, 10 kilometres per hour). But the tortoise is still ahead, if only by one-tenth of a kilometre (the distance travelled is the speed, one kilometre per hour, multiplied by the time, one-tenth of an hour, or six minutes). Continuing on in this way, Achilles runs up to the point the tortoise last was, only to find the tortoise is still ahead. The reasoning seems to go on forever, so Achilles never catches up. But of course he does, as we see that after two hours he has passed the tortoise. So what gives?

The answer is that infinitely many small amounts do add up. The distances that Achilles travels trying to catch up to the tortoise are, in kilometres, 1, 1/10, 1/100, 1/1,000, ... If we call the sum of all of these S, then $S=1+1/10+1/100+1/1,000+. . . .$

We now pull a little trick of multiplying both sides by 10: $10S=10+1+1/10+1/100+. . . .$ Now we notice that the right-hand side is almost the same as before, except the second example starts with 10 rather than 1. By subtracting the second from the first, we get almost everything on the right to cancel:

$$10S{-}S = 10$$
$$9S = 10$$
$$S = 10/9=1.11111. . . .$$

Thus Achilles does catch up to the tortoise, at the 10/9 kilometre point, or 1.111 . . . . kilometres. After that point, Achilles is, of course, always ahead.

Now I hid a subtlety in that I subtracted two infinite amounts and did some cancelling, but the bottom line is that many small infinite amounts add up (and that Achilles can still take great pride in being able to beat a tortoise in a race).

INFINITE SUMS

Adding up infinitely many numbers is something that can be infinitely fascinating. The notion of infinity is a truly mind-bending concept for which there is no equivalent in real life. I will never experience infinity (though sitting through some math lectures I have approached it).

Infinity is a concept that has its foundations in a time long before mathematics made it precise. Much of monotheism is based on the concept of a G–d that is infinite, that has no beginning or end. Monotheism requires the concept of infinity, the whole being the sum of its infinite parts.

One of the nice things about mathematics is that there are often many ways to visualize proofs. For example, what, if anything, does 1/2+1/4+1/8+ . . . add up to? I like to think of this as a 1×1 square, cut up into pieces.

The first cut divides the square in half, and has area 1/2 units.

The next cut divides the remaining white square in half, and this piece has half the area remaining, that is, 1/4 units.

We continue this process. Eventually, the entire square becomes black. "Eventually" here means after infinitely many steps. The sum of the areas of all the black pieces we remove, which is 1/2+1/4+1/8+ . . . , will be 1, the area of the whole square. So we have a "picture proof" that in fact 1/2+1/4+1/8+ . . . =1.

Many non-mathematicians think of mathematics as symbol-pushing, especially when it comes to proofs. Proofs may appear to be lines of mathematical calculations, with what you want to prove somehow appearing magically at the end. But this is far from the truth. Almost all mathematical proofs, at least the best parts, start off as a mix of pictures, experiences, and intuition. It's only at the final steps that mathematics erases much of the path taken to provide a pristine argument. I personally like the visual, error-prone, sweat-filled, human work that occupies me and other mathematicians.

THE TAIL WAGGING THE DOG

Now, the long tails of distributions are just a large set of smaller and smaller numbers that can add up to significant amounts.

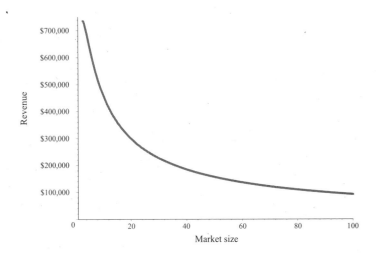

Above is an example of what the revenue might look like for a company, with the size of the markets listed along the bottom. So, for example, the number 20 along the horizontal axis might represent the twentieth-largest market, with the height on the curve at that value being the revenue that that market can generate, which looks to be about $280,000.

The company might think that it is not worth considering any market that generates a revenue below $200,000. This would be from about the thirty-sixth-largest market onwards. But what happens if you add up the demand from the thirty-sixth-largest market, until, say, the hundredth-largest market? It adds up to about $83 million plus change! This is certainly worth considering. With the Internet making such markets easily accessible, there is a lot of money to be earned from the long tail.

COUNTERINTUITIVE COUNTEREXAMPLES

The mathematician in me focuses in on the addition of infinitely many numbers, rather than the money to be made from the long tail. It's sad, isn't it? (My wife thinks so.) In any case, it seems reasonable that infinitely many smaller and smaller positive numbers always add up to a number, but such is not the case.

Finding examples like this is the other side of the "proof" coin for mathematicians. If a general statement is true, mathematicians need to hunt down a proof. But if it's not true, all that needs to be found is a *counterexample* – a single example to show that the statement can sometimes be false. It might sound like this would be easy to find, but often it is like searching for a needle in a haystack (an infinite haystack, at that). And on top of it all, you often don't *know* whether a statement is true (and you should look for a proof) or whether it is false (and you should look for a counterexample). Many a mathematician has spent years, if not a lifetime, looking for the wrong answer.

So what about the statement that every set of smaller and smaller numbers adds up to a number? Well, think about adding up the reciprocals of all the positive integers: $1/1+1/2+1/3+1/4+1/5+1/6+1/7+1/8+\ldots$ This is known as the *harmonic series*. Surely, the harmonic series must add up to something, right? Well, I've tried using my computer to add up the first hundred of these, and I get a little less than 5.2. Adding up the first thousand terms gives a little less than 7.5. And adding up the first 10,000 terms gives a little less than 9.8. Perhaps a good guess is that the numbers add up to 10? It would be a pretty good guess except that it's *dead wrong*!

Here is the argument (or proof) that the sum is infinity: that is, the sum gets as big as you would like. The third and fourth terms, $1/3$ and $1/4$, are each at least as big as the fourth term, $1/4$. Likewise, the fifth through eighth terms are all at least as big as the eighth term, which is 1/8. The next eight terms are at least as large as the last term, 1/16, and so on. Thus we have that

$$\frac{1}{1}+\frac{1}{2}+\frac{1}{3}+\frac{1}{4}+\frac{1}{5}+\frac{1}{6}+\frac{1}{7}+\frac{1}{8}+\ldots \geq \frac{1}{1}+\frac{1}{2}+\frac{1}{4}+\frac{1}{4}+\frac{1}{8}+\frac{1}{8}+\frac{1}{8}+\frac{1}{8}+\ldots$$

$$= \frac{1}{1}+\frac{1}{2}+2\left(\frac{1}{4}\right)+4\left(\frac{1}{8}\right)+\ldots$$

$$= 1+\frac{1}{2}+\frac{1}{2}+\frac{1}{2}+\ldots$$

and the last bit gets as large as we like (by adding say one hundred 1/2's we get 50, and so on).

The surprising thing is how slowly this series (the sum of infinitely many numbers) grows. We've seen that it takes more than 10,000 terms to get close to 10. If you add up the first billion terms, you'd only get as far as 21 and a bit. Pick any positive number and eventually, by adding up reciprocals of positive integers, you will get a larger number than the number you picked. If you decided to add up the terms in the harmonic series, one every second, from age 20 until age 120, without sleeping or stopping, you wouldn't even reach 23.

And if that isn't bad enough, there are series that go off into infinity, but at an even slower pace! One that does is the sum of the reciprocals of all the prime numbers: 1/2+1/3+1/5+1/7+1/11+ . . . It's not obvious that this sum gets arbitrarily large (the eighteenth-century mathematician Leonhard Euler was the first to show that it does). If I add up the reciprocals of the first 100,000 primes (the 100,000th prime is 1,299,709), I only get to about 2.906; if I add up the reciprocals of the first 100,000 positive integers, I would get to about 12.09. The series of the reciprocals of all the primes adds up very, very slowly, and it's beginning to put me to sleep.

One fascinating application of harmonic series is to record-breaking, the tracking of values that can go up or down over time. Suppose you have some value that you measure from time to time; it could be the average goals-against in the NHL; the fastest time in the 100-metre backstroke at the Olympics; or the annual rainfall in your city. For any of these, the expected number of record breakings, just due to chance, over n contests turns out to be the sum of the harmonic series up to $1/n$, that is, $1/1+1/2+1/3+\ldots+1/n$. If you notice that there are significantly more or fewer record-breaking instances, you can be fairly sure that something other than chance is at play.

One thing I have just tried is to look at some of the data for global warming. On the U.S. National Climatic Data Center's global warming site I found a graph of the average land temperatures from 1880 to 2001 (the bottom bar graph). The graph shows the annual average temperature compared to the overall average temperature (at the 0.0 line), allowing us to read off the number of record-breaking years. Of course, the first year is considered a record breaker (as there are no previous records). I then count the number of times a bar of the graph was higher than all previous bars; these are the record-breaking years. I count 14 record-breaking years in the 122-year period from 1880 to 2001 (remembering to count the years 1880 and 2001). How many should I expect just by chance? I would expect $1/1+1/2+1/3+\ldots+1/122 \approx 5.39$ record-breaking years, so we've well exceeded this number.

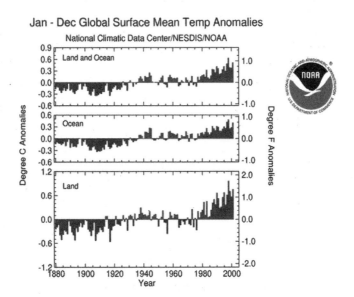

The harmonic series grows very slowly, and so we don't expect the number of record-breaking years to grow quickly unless some non-random factor is at work. Whether it's global warming or some other factor, I am convinced that chance cannot account for the discrepancy.

YOU DON'T KNOW THE HALF OF IT

To truly understand series, you need to get down to the basics of what the sum of an infinite series means, which is not at all obvious. In fact, many great mathematicians worked with infinite series long before the details were ironed out. Students now learn about limits and infinite sums in first-year calculus, but it wasn't so long ago that infinite sums baffled many of the world's best mathematical minds.

Most of the discussions revolved around what a *limit* meant. In the 1850s mathematician Karl Weierstrass expounded the epsilon-delta definition (which had been introduced earlier by another mathematician, Bernard Bolzano, in 1817, but it didn't catch on). The definition is quite abstract, and only lucky (!) calculus students get to see it in first year. The epsilon (ε) and delta (δ) are lowercase Greek letters used to denote arbitrarily small positive amounts.

Now we get back to infinite series. Infinite series are made up of a first term, a second term, a third term, and so on that we want to add up. In the generality of mathematics, we give names to everything, and a good naming scheme is worth a lot. So let's call the first number in our list a_1, the second a_2, and so on; the ith number in the list will be a_i. So what we want is $a_1+a_2+ \ldots +a_i+ \ldots$ Rather than write the dots, mathematicians abbreviate this with what is called *sigma notation*, after the Greek letter sigma (Σ), which is used to stand for "sum." The infinite sum $a_1+a_2+ \ldots +a_i+ \ldots$ is written as $\sum_{i=1}^{\infty} a_i$; both mean the same thing, but the latter is more compact and tidier to write. It looks more complicated than it really is. Now in detailed terms, mathematicians say that an infinite sum $\sum_{i=1}^{\infty} a_i$ equals some number L if you can make the result of adding up the first so many terms as close as you like to L. In even more detail, whenever you choose any small positive epsilon ahead of time (for example, say 0.0001) you can *always* make sure the sum of the series is within epsilon of *L, whenever you add up enough terms in the series.* So for example, suppose you want to show that $1/2+1/4+1/8+ \ldots =1$. If we take epsilon to be 0.0001, we need to make sure that with enough terms, the sum is within 0.0001 of 1. Now there happens to be a formula for adding up the first n terms in the series, which you can discover by using the same trick I used for adding up Achilles' distances in the race against the tortoise: $1/2+1/4+1/8+ \ldots +(1/2)^n=1-(1/2)^n$. This is only about adding up a finite number of terms, so is more straightforward than adding up infinitely many terms. Now provided that n is large enough, that is, provided that n is at least 14, then this value is within 0.0001 of 1, as $1-(1-(1/2)^n)=(1/2)^n \leq (1/2)^{14}=0.000061. \ldots$

Now to prove that the sum is exactly 1, you'd have to prove that you get within epsilon of 1, no matter how small you choose epsilon to be. Does this mean infinitely many proofs? No. What you have to do is handle all such proofs within one argument. It's not much harder than what I've written down in the specific case of epsilon being 0.0001, but I'm worried that I'm beginning to wear out my welcome.

A BEAUTIFUL MIND

All this talk about epsilons reminds me of my undergraduate days in Calgary, and the start of my love affair with mathematical research. During my second year, my calculus professor suggested I apply for a summer NSERC grant (the Natural Science and Engineering Research Council is the government arm that supports research in the sciences). Research in mathematics? It was something I hadn't really thought about. The professor suggested I look at graphs, and I pictured the kind of plots I had seen and drawn in calculus. What could be new or interesting about those?

I soon discovered I was way off-base. My prof dropped a classic graph theory book on me (it only hurt for a short time); I worked through it that summer and fell in love with graphs and networks. I found that I could indeed find brand new results.

In the early eighties the mathematics department at the University of Calgary was a vortex of activity in discrete mathematics (the area of mathematics that includes graphs). It seemed that there were several seminars each week in discrete mathematics, and many in the department were busy researching the area. Even those whose area of research lay outside discrete mathematics were drawn toward it in weird and wonderful ways.

There was, perhaps, no one better at seeing such connections than Paul Erdös, whom I had the privilege of meeting. Erdös (1913–1996) was a Hungarian mathematician who roamed the world, visiting one university after another, doing mathematical research wherever he went. He was arguably one of the most famous mathematicians of his time. Erdös's eccentricities were part of his charm. Until he was in his seventies, his mother travelled with him. He often didn't take suitcases, using shopping bags instead. And he had a unique language of his own. He called children "epsilons," because they were so small.

Paul Erdös was revered when he visited Calgary's mathematics department, and one day he chose to take me, a mere epsilon at the time, out to lunch. I knew of his work and, awestruck, I was fairly quiet at lunch. He drew me out, and it was wonderful to speak with him about mathematics, and also about life in general. He wanted to go to a nearby plaza to get his hair cut, so I happily walked with him. The intersection for the plaza was always incredibly busy and, like everything in Calgary, very large. Abstract concepts in mathematics obviously were very concrete to Erdös, but he seemed to consider concrete objects like traffic lights and cars and trucks travelling at high speeds to be abstractions. As he strode across traffic intersections against the lights, I trailed behind with a prayer. I was terrified of being the young mathematician who would forever be known as the one who was with Paul Erdös when he died crossing the road.

A final episode filled that day. Erdös asked me to accompany him to the library to find an old paper of his that would help him recall the details of a result. He had an idea of what journal and volume number it was, and I went off to fetch it while he browsed the shelves. The series numbers for the journal stopped before the number he had suggested, and when I returned empty-handed, he grew impatient. I said nothing, but when Erdös checked the stacks himself with the same result, he apologized to me. The great Paul Erdös apologized to *me*, a lowly epsilon. That day I learned that a great researcher can be an even greater *mensch*.

TRIMMING THE (GOVERNMENT) FAT

This discussion of adding up infinitely small numbers may sound esoteric, but similar situations do crop up in life. Consider this riddle:

> Question: When do infinitely many zeros add up to something huge?
>
> Answer: When the government is involved!

Let me illustrate how numbers can be used to trick the unsuspecting public. While trying to cut down on dietary fats in general, I am particularly staying away from trans fats. From what I've read, there is no safe level of trans fats – any is too much. So I have declared my body a trans-fat–free zone and read food labels very carefully. A couple of months ago I was eating my morning bowl of cereal, which I knew was free of trans fats, thanks to the nutrition label, which said "0 g trans fats per serving." But my wife pointed out that the box listed hydrogenated vegetable oil as an ingredient, which is a trans fat! So I stomped to the telephone and called the company, whose representative told me that, indeed, there are trans fats in the cereal. The Canadian government allows companies to round any amount less than 0.2 grams of trans fats per serving down to zero! So if, say, there are 0.19 g of trans fats per serving, the manufacturer can round down to zero and claim their product is trans-fat–free. Over days, months, and years, you could be ingesting a significant amount of artery-clogging trans fats, all the while believing you are conscientiously staying away from the stuff!

Here is the kicker. Even if I were okay with this rounding business (which I am not, on both health and mathematical principles), at least it seems to make a level playing field for all companies because it's always per serving. But I have found out that *there are no guidelines for what constitutes a serving* for a product. Each company can set its own serving size.

Go and look at the nutrition labels for some of your favourite chocolate bars, and check out the listed serving size. You will be surprised to see that often the serving size refers only to a part of the bar. Who on earth would think that a serving of a chocolate bar would be half of the bar?

I'm looking at a can of salmon. The can contains 170 g, but the serving size is 100 g. I could understand a serving being half of a can, or 85 g, but I can't imagine sitting down and saying, "Hmm, I'd like a serving of salmon. I think I'll have ten-seventeenths of a can." Can you?

Having looked at food labels on many canned and packaged foods, I have come to the belief that some (but not all) companies choose their serving sizes by the amounts that give them the best rounding. For example, I can imagine the following scenario. There is a pouch (which is a serving) containing 100 g of food with 0.2 g of trans fats, which the company would have to declare. But if they instead put 95 g of the mixture in the pouch, the amount of trans fats falls to 0.19, and now the company can round this amount down to zero and claim that the product is free of trans fats!

The moral of the story is that you have to look beyond the nutrition facts table to the ingredients, and be wary of how mathematics can be used to put a falsely healthy face on some products. Moreover, what you believe is a sum of zeros, namely the amount of trans fats listed for what you are eating, can eventually add up to a significant amount of trans fats, when the true amount in each portion is some small but positive value. No matter what you learned in school, sometimes zero plus zero isn't zero.

Break's over; back to work. As I stick my head back into the stack of paper, I think about how many, many small events add up to a lifetime. Each moment accounts for so little, and yet the sum is so significant. All we can all hope for is a very long tail at the end.

NETWORKS, FAME, AND COINCIDENCES

I've just received yet another e-mail announcing a mathematics conference, this one in Hungary. A math conference is an opportunity to get together with fellow mathematicians I haven't seen in years, to talk research – and to catch up on sleep during the lectures.

Conferences are always interesting social occasions. We mathematicians are a bit of a strange lot. When I was an undergraduate at the University of Calgary, I remember that the prevailing shoe style in the math department was sandals with white socks. My wife finds mathematicians' grooming – or lack thereof – greatly entertaining. To be certain, there is always an interesting array of facial hair: eyebrows used as comb-overs, nose hairs long enough to braid, and, my favourite, natural ear muffs. Luckily for me, my wife takes care of these things, so I don't scare off too many people.

On the plane on my way to a recent conference at Vanderbilt University in Nashville, I chatted with the woman sitting next to me. I've always been a bit shy, but I knew that if we talked long enough, I was likely to find some coincidence to connect us, and this helped me get past my shyness. We might know someone from the same high school, share a common friend, or I might be in contact with a relative of hers: the "six degrees of separation" effect. Provided I looked hard enough, it was likely I would find some connection. Comedian Gilda Radner was right – it's always something!

We all have our own social networks. My wife has always been close to her large extended family. She knows her third, fourth, and fifth cousins well, and she has many, many friends all across North America and beyond. Her social network is extensive and, for me, a little overwhelming. When she took me to Winnipeg to meet "everyone," little did I grasp how close "everyone" was to its true meaning. Like many mathematicians, I am not normally good at remembering names. If only people had numbers: "Hello, my name is 65, the smallest odd number that can be written in two different ways as the sum of squares." But no, everyone has to have a name like "Susan" or "Mallory." Anyway, I felt like royalty, as I greeted all of my wife's numerous friends and family.

I, on the other hand, grew up among a much smaller circle of relatives. Pretty much just first cousins, with a few second cousins thrown in for good measure. My friends over the years were a relatively small set, too. In each period of my life I had a couple of close friends, with a few more on the periphery.

I have had to fight against my innate shyness my whole life. We mathematicians aren't known for our social skills. I have read recently that there is a correlation between mathematical ability and scoring high on the autism scale. This doesn't surprise me. Much of mathematical reasoning is a solitary, internal pastime, and the more you can exclude the external world while you are reasoning the better. I learned early in a big household to be able to concentrate in noisy environments. I am still able to multitask and read a book or do research while watching television.

Some social networks are small, and some are large, and they change over time. My wife and I now find ourselves living in a community where most of the people are "bluebloods," having lived on the east coast for a number of generations, and we are the CFAs – the "come-from-aways."

I think of social relationships as a graph or network. In this view, each of us is a point. Two people are joined by a line if they know each other. This is a *big* graph if you want to include everyone in the world, one with more than 6 billion points – not one even I would want to have to draw. And, of course, the graph changes from moment to moment.

What is absolutely fascinating is that there *is* some structure in such a huge social network. The structure has been coined *six degrees of separation*, and it goes as follows. Between any two people in the world's social network, there is a chain of at most six lines that connects them. That means that for any random person I meet, it is likely that there will be some chain like: I → have a friend from high school → who has a cousin → who went to university with someone → who knows your sister's best friend → who knows your sister → who knows you. Often the chain is shorter, but in practice this "six degrees of separation" has been observed as a general rule. A game known as Six Degrees of Kevin

Bacon attempts to connect a certain actor to Kevin Bacon in as short a chain as possible by using someone who has appeared in a film with Bacon as the first link and tracing the connections back to the target actor. (Apparently Kevin Bacon once said that if there was an actor he hadn't worked with, then there was someone who he had worked with who had worked with that actor.)

On the one hand, it may seem incredible that somehow, in just a few steps, you can be connected to anyone in the world. Just on the basis of numbers you might think that it's impossible, as you probably know at most only a few hundred people, and there are *billions* of people in the world. But it's not beyond the realm of possibility for the following reason. Suppose you know 300 people. And suppose each person you know knows 300 people (so those three hundred people as a group have $300 \times 300 = 300^2$, that is, 90,000 acquaintances). Those 90,000 people know $300 \times 90,000 = 27$ million people, and those 27 million know 8.1 billion people – within only four steps of knowing you! This is how quantities can explode *exponentially* when you are multiplying at each stage. Certainly not everyone knows exactly 300 people, and we've overcounted people, as some of the people you know know each other, and so on (there aren't 8 billion people in the world yet, so the approximation can't be completely correct). But you get the idea that it is possible, if perhaps unlikely.

Now of course no one has proven the rule. To do that, you would have to find out what are the social connections between all people in the world, a daunting task. And the network is always changing, with people being born, dying, and making new connections. The original six degrees of separation principle was conceived in an experiment by the psychologist Stanley Milgram, who randomly chose people in various cities in the United States and asked them to send a letter to another person in a different city. If they knew the person, they could send it directly; otherwise, they could send it to a friend of theirs who they thought would best be able to forward the letter onwards. What was observed was that, on average, the letter took at most six trips to reach its final destination. Of course, the letter-sending experiment had no way of checking whether the letters were being sent by the shortest path possible (as people could only guess at who might know the recipient better).

The principle Milgram discovered is sometimes also called the "small-world phenomenon." It is quite surprising, I think. The world consists of billions of people. We each know personally only relatively few people – perhaps a few hundred (in my case) or a few thousand (in my wife's case). While our friendships tend to be clustered (most of our friends are friends with each other, so there is a lot of overlap between who we know and who our friends know), we each also know a few people outside our cluster of friends – perhaps someone who now lives in Japan, or who married someone from Europe. It is these seemingly random, jumping links that bring the world closer together. Ultimately, we are not so far apart. There is no guarantee that you and a random person can find a chain of six or fewer connections joining you, but it happens so frequently that it's worth pursuing with new people you meet. They are bound to be amazed by the connections you discover.

Again, being a bit mathematically inclined has an advantage: it lets you see coincidences for what they are – just coincidences. I remember one time I was driving and I passed by the house of someone I knew, a man named David Tax. Right at that point a song played on my CD player, Stevie Ray Vaughan singing George Harrison's "Taxman." Was it a sign? Probably not. It *was* a strange coincidence, but such things are just bound to happen, by chance. We humans are inordinately good at looking for meaning in everything, even when it's not there. We just don't pay attention to all the times when coincidences don't occur.

MAPPING OUT FRIENDSHIPS

In my mind, I envision the relationship of who-knows-whom as a huge network; each of us is a single point, with lines joining two points if the people know each other personally. For a small group of people I could probably spend the time to find exactly who knows whom, but the true world network is something beyond my ability to map out. Instead I sketch out an abstract picture, something like the following:

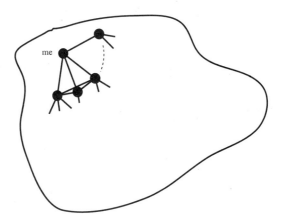

I imagine that the picture has many points looming, with many lines joining various pairs. The dotted line indicates that there are untold numbers of other points to be filled in. To this, I might add a chain to show that there will be at most six lines between any two points.

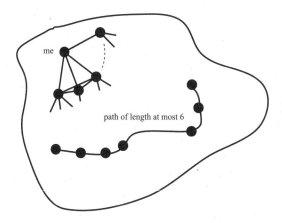

This picture seems to cover all of the relevant aspects of the world's social network.

PIGEONHOLING PEOPLE

While the six degrees of separation principle is *almost* certain to work, there is another mathematical principle for which there is complete certainty. It's the *pigeonhole principle* I mentioned in Chapter 4. It goes as follows. Suppose you have more pigeons than pigeonholes. Then no matter how you put the pigeons into the pigeonholes, there will a pigeonhole that has at least two birds in it, guaranteed. In fact, if there are n pigeons and m pigeonholes, there must be a pigeonhole with at least the average number of pigeons (n/m pigeons) in it.

The pigeonhole principle is rather obvious, but incredibly useful. The interesting thing about it is that it is what is called "nonconstructive" – you get a conclusion about something happening for which there is no general procedure for determining how it happens. You know that there is a pigeonhole with more than one pigeon, but the principle does not tell you which pigeonhole it is. In Chapter 6 I talked briefly about the mean value theorem, which tells you that you must at some point in a trip have achieved your average speed over the trip, but does not tell you at what point this occurred.

There are many uses for the pigeonhole principle, depending on how you interpret the pigeons and the pigeonholes. For example, if I have a dryer full of socks, of which there are, say, five colours (red, black, green, navy, and brown), and I pull out at least six socks, I am guaranteed to have a pair. The pigeons are the socks and the pigeonholes are the colours. The pigeonhole principle doesn't tell me which coloured pair I will get, only that I am guaranteed to get a pair. If I pull out five or fewer socks, I might get a pair, but then again I might be unlucky and get only singletons.

Now you might think that a principle as simple as the pigeonhole principle isn't of much use. Quite the contrary. When you write any rational number (that is, the division of any two natural numbers) as a decimal, it has to repeat after a certain point. You may have learned this in school, but you probably don't know why it's true. It's the pigeonhole principle at work. Let's divide 85 by 62. It starts off like this:

$$62\overline{)85}$$

Here are the first few steps:

$$\begin{array}{r} 1 \\ 62\overline{)85.0000000} \\ \underline{62} \\ 23 \end{array}$$

$$\begin{array}{r} 1 \\ 62\overline{)85.0000000} \\ \underline{62}\downarrow \\ 230 \end{array}$$

$$\begin{array}{r} 1.3 \\ 62\overline{)85.0000000} \\ \underline{62}\downarrow \\ 230 \\ \underline{186} \\ 44 \end{array}$$

The numbers that you get by subtracting (23, 44, and so on) are all remainders when you divide by 62, and are therefore between 0 and 61 (otherwise, you haven't divided 62 properly into the number). Under the pigeonhole principle, by the time you have done at most 63 steps, you must repeat a number (there are only 62 "pigeonholes," namely 0, 1, 2, 3 . . . 61, and the numbers you get by subtracting down are the pigeons). And once you repeat a number, the whole process repeats from that number onwards, so the decimal expansion of the rational number 85/62 must repeat.

There is nothing special about the number 85/62; the same principle holds true for the division of any two natural numbers, that is, of any rational number – as a decimal, it *must* repeat (or terminate with infinitely many zeros, which is still a repetition).

Here is another rule about social networks, one that, unlike the six degrees of separation principle, is bound to hold, no matter what. Let's call a group of three people, each pair of whom have met before, a group of three "mutual acquaintances." And let's call a group of three people, no two of whom have ever met before, a group of three "mutual strangers."

Now if I walk into a given room, I don't know who has met whom. But I am guaranteed of one thing: if there are at least six people in the room, then either there is a group of three mutual acquaintances or a group of three mutual strangers among them.

Certainly, if in the room of at least six people every pair of people has met before, there is a group of three mutual acquaintances, and similarly, if no pair of people have met before, there is certainly a group of three mutual strangers. But what about in all the other numerous cases where some people have met and some haven't?

Here is the argument. Pick six people in the room and focus on one of them; let's call her Alice (she doesn't live here anymore).

Alice

Among the other five people, Alice has either met at least three of them or there are at least three of them whom she hasn't met.

(Think of two pigeonholes, one labelled "Has Met Alice" and the other "Hasn't Met Alice," and put each of the five people in the correct pigeonhole. Then one of the pigeonholes must have at least the average, namely $5/2 = 2\ 1/2$ pigeons. But you can't have a fractional number of pigeons, so that pigeonhole must have at least three pigeons in it.)

So now there are two cases. If Alice has met three people, and if any two of these have met before, those two along with Alice are a group of three mutual acquaintances; otherwise, no pair of Alice's acquaintances has met each other, so three of them form a group of three mutual strangers.

ALICE HAS MET THREE PEOPLE

The other case, where Alice has not met at least three of the other people, can be handled in exactly the same way.

ALICE HASN'T MET THREE PEOPLE

IT'S A BIG WORLD AFTER ALL, THOUGH NOT AS BIG AS YOU MIGHT THINK

A decade ago, I would never have envisioned that the connections in the World Wide Web would begin to dwarf the connections among people. But it has happened. I read somewhere that there are now an estimated 20 billion web pages!

The links among web pages has been the focus of study, much as social networks have. Both display inherent structure, in spite of what seem like random changes over time. One of the things that mathematicians do is seek commonality among apparently different-looking structures. It has been observed that the link structure among web pages has the small-world property, in that for any pair of web pages, there is a short trail of links to travel from one to the other. The Web in fact depends on such links – without them it would be too cumbersome to navigate.

But the Web, like a social network, has no one overseeing it. It self-organizes. What is amazing is that it does so on an ongoing basis. What accounts for this? Well, one idea put forward is that while the Web has many pages, there are relatively few pages that are of high importance. Some of these are called *hubs*, which link to a large number of web pages, and *authorities*, to which a lot of web pages link. A web page to which a lot of web pages link is likely to be a good reference on a topic, and hence the name "authority." Likewise, hubs are likely to be pages that point out other pages with a lot of good references on a topic.

Turning the connection backward, these ideas shed light on why social networks have the small-world property – there are also people who have an inordinately large number of acquaintances. These "hubs" bridge the gap between everyone's clique of friends and the rest of the world.

There is a principle that has been put forward that the web pages that have many links coming in to them tend to attract more inward-pointing links, as they are so popular. In terms of friendship graphs, it states that we tend to become friends with people who have lots of friends. This principle is known as the "rich-get-richer" factor and it has been shown to be able to account for small-world network properties of hubs and authorities.

I just read an interesting article about slowing the spread of sexually transmitted diseases. The article centres on the application of the small-world property to sexual relationships. The ease with which sexually transmitted diseases are spread is often due to the fact that among a significant part of the population the sexual distance between many pairs of people is rather small, and hence it doesn't take many encounters for an STD to become an epidemic.

What I find interesting is that the small-world aspect is likely due to the presence of "hubs" or, to put it delicately, particularly active individuals. To control the spread of a disease it would be most worthwhile to focus in on those who are particularly active and convince them to curtail their activities. Such an approach would have greater benefit than spreading limited resources more thinly among the whole population.

GOOGLE ME!

I am very impressed with Google's searching capabilities. Just type in a few keywords, and off you go into cyberspace. Within seconds you are given a list of relevant web pages, to a large extent in decreasing order of relevance. There are *billions* of pages for Google's search engine to deal with, and the response is so quick, and so accurate.

I naturally did a bit of searching about Google's searching capabilities, and found that it all depends on mathematics. The idea that the founders of Google had was that their search engine would have to rank web pages on the fly, according to whatever keywords were requested. The rank of each page would depend on a general ranking of the pages that is more or less static, and then some minor modifications of the ranks for the particular keywords chosen. Their thought, for the static part of the page ranks, was that the ranks should reflect the average of the page ranks of all of the pages linking *into* the page (rightly, I think, the procedure ignores links *from* a page when assigning it a rank, as anyone can create links out of their own web page).

To get the page ranks for all the web pages would require a *huge* calculation, well beyond the capabilities of present-day computers. But a bit of math lets you start with any assignment of page ranks to each page, and then place the page ranks into a machine that updates them. This final set of page ranks would be the static part of the page ranking. All of this depends on something called the *eigenvalues* of matrices. Imagine you have a rule that takes any number, halves it, and then adds 5 (that is, the rule is the function $f(x) = x/2+5$). One thing a mathematician would be interested in is: what numbers does the rule fix, that is, not move? On the one hand, you can go off and try to solve for it (it's not too hard). On the other hand, see what happens if we simply keep applying the rule to *any* number. If you take, say, 32, you would get $(32/2)+5=21$. Then 21 gives you $(21/2)+5=15.5$, which gives you $(15.5/2)+5=12.75$, and so on. After 10 times you would be at 10.02 . . . and subsequently would be getting closer and closer to 10. And it wouldn't matter what number you started with. If you started with 1,492, it would take you a little longer, say 20 steps, but you'd be getting close to 10 as well (after 20 steps the number is approximately 10.0014). So if it was 10 you were after, you could just run the procedure on any starting number and wait for it to settle down. The place it settles down to is called an *attractive fixed point*. (It's attractive because it attracts the other numbers like a magnet, not because it's prettier than the rest.)

The page ranks are very much like this, only there is a whole set of numbers that we want as the page rank, not just a single one. The solution is a fixed point, and while it is a straightforward calculation, not all the processing power in the world could solve it. The repeated procedure that hones in on the solution is something that processing power (a fair bit, I might add) can handle. An approximation is sometimes better than the real answer (especially when the real answer costs too much).

MINING FOR GOLD

What Google excels at is mining the huge amount of data on the Web for golden nuggets of information. Information is a valuable commodity these days. There is a whole branch of computer science called data mining that explores ways to get useful information from reams of data. And yet often data collection is done surreptitiously.

I am not paranoid; I *know* I'm being watched. Companies are monitoring my every buying habit, my every purchasing tendency. There are all sorts of mathematical tools that are being used to pry dollar bills from my wallet. Some of these take their inspiration from human biology. There are networks modelled on our neurological system that learn to make better and better predictions by training on data, and there are genetic algorithms that are based on the concepts of evolution and DNA recombination.

What these companies all are after are patterns, or trends, in buying, for individuals and for the general population. Once they isolate the patterns, they can modify their approaches and spend their resources on the individuals who are most likely to make a certain purchase. Mathematics is the ideal tool to look for patterns wherever they may be hidden.

A VOTE OF CONFIDENCE

Politicians are crazy for patterns, especially among the electorate. The mathematics of voting is easy to understand, but some of the paradoxes are not. Suppose that we have three people (Al, Bob, and Cynthia) willing to serve as president of our community group, PWLMATA (People Who Love Mathematicians and Accept Their Absentmindedness). Suppose everyone plans to rank them, from top to bottom. For instance one person might rank Bob at the top, Cynthia next, and Al at the bottom. Suppose when you gather all of the lists from the group's 100 members, you get the following results (from top to bottom in each case):

- 30 people voted: Al, Bob, Cynthia
- 35 people voted: Bob, Cynthia, Al
- 35 people voted: Cynthia, Al, Bob

Who should win? Well, if you compare Al and Bob, Al ranked higher than Bob by 30+35=65 voters (those in the first and last blocks), so Al should beat Bob. By the same reasoning, Cynthia beats Al 70 to 30, an even bigger margin. So Cynthia should win, right? Well, if you compare Cynthia and Bob, you find that Bob wins, 65 to 35.

Mathematics has proven that there is really no general process for combining voting preferences. Suppose that in your community group you want to rank a list of options. They might be a list of 10 requests for money from the parent-teacher association. You might ask everyone in the PTA to rank the 10 options, from 1 to 10. How should you combine the lists? If what we're after is a general principle, we'd want a certain number of rules to hold. For example, we wouldn't want a dictator, someone whose list always determines the group's ranking. Secondly, adding new options shouldn't alter how we rank the other items with respect to one another. Thirdly, if everyone prefers option A to option B, then the group as a whole should prefer A to B as well.

Back in 1951 an economist, Kenneth Arrow, proved that if you accept just these three reasonable rules and there are at least three options to rank, there can be no general procedure for making a group ranking. It isn't just a matter of not being smart enough to think of a procedure to rank items for a group – *there can't be such a procedure!*

I love these *impossibility theorems*. They are just so counter-intuitive that it's hard to get your head around them. Such theorems demonstrate that even mathematics has its limits.

Now what does this say about all those important things we vote on, like political leaders? Well, first of all, we all cast only a single vote, rather than ranking each of the politicians. So we aren't after a group ranking from individual rankings. Also, there are many people who would find good reason to say that the current voting system in Canada isn't fair. It is not a proportional system; not everyone's votes count the same. And don't get me started about the Senate, our unelected second House.

My talk at the conference in Nashville was about the many connections between mathematics and music. I was determined to put on a good show, and it was my first research talk where I had my electric guitar strapped to me. At the Nashville airport on the way out, there was a red-haired lanky man with a laptop, sitting on the floor of the terminal. I instantly recognized him. Shyness or no shyness, I wasn't going to miss this opportunity and I struck up a conversation with the Barenaked Ladies' bass player, Jim Creeggan, about music and mathematics. Instantly, my "degree of Jim Creeggan" dropped to one, but to be fair, Jim Creeggan's "degree of Jason Brown" dropped down too!

10

INTERNET BANKING
AT PRIME

I'm home for the evening now, and before dinner, decide to pay some bills online. It took me a little while to get comfortable with doing my banking on the computer. I am still leery inserting my credit card number on online forms – I have no idea where the data goes, who has access to it, and how secure it is.

I know that data encryption, when used correctly, can safeguard my data – provided no future number-theory genius uses his powers for evil rather than good.

Money is arithmetic in action. We all learn to do basic mathematics in order to keep track of our cash. Saving money, making money, spending money, borrowing money – these all require some elementary arithmetical skills. In fact, the concept of negative numbers is often taught in the context of borrowing money.

In a popular *Saturday Night Live* sketch a bank commercial spends the entire time explaining how the bank can accurately make change for you. For a $20 bill they will give you two $10, or four $5, or one $10 and two $5, and so on. The theory of partitions of whole numbers in action!

MONKEY BUSINESS

My parents were in business for many years, running a successful accounting company for mining firms, on Bay Street in Toronto. Somehow I never thought of going into business, though I loved spending summers in Toronto, running errands all over the city for my parents. I wish I had learned more from them. But once I was infected with the math research virus, there was no turning back.

Today in academia, I do teach budding business people. One of the big topics I teach is something called *linear programming*, which involves solving many large-scale problems for maximizing profit or minimizing cost, when you have limited resources.

Sometimes we university professors get lessons in finance from our students. I remember one fellow math professor who had taken it upon himself to mentor gifted high school students in university mathematics. My colleague had been teaching one particularly bright student some of the prettiest mathematics, and asked him what he wanted to be when he grew up. The student replied, "A pharmacist." The professor tried his best to promote being a university mathematics professor as the ideal profession for one so gifted. In response the student asked the professor, "How much do you earn a year?" The professor, taken aback, replied with an amount. The student quickly told him that it was a pittance compared with what he would earn as a pharmacist. The discussion ended with a long period of silence.

Of course, there's more to life than money! At least that's what I repeat to myself. But if you are monetarily motivated, you need to know some mathematics. It's a shame that so few people do, as a little more knowledge will keep many more dollars in your pocket.

FOR INTEREST'S SAKE

How much interest do I earn on my money in the bank? How much do I pay out in interest on loans and on credit cards? All of these questions are answerable, using mathematics. I think that most people don't realize how the calculations are done, and simply put their trust in the banks. Of course, why not trust them, as they have your best interests in mind? Or do they? Luckily, the mathematics of money isn't really that difficult.

Let's take an example. Suppose I have deposited $12,000 in an interest-bearing account, with the interest rate at 6%, calculated monthly. What this means is that the interest rate per month is one-twelfth of the total interest rate, 6%, which is a yearly interest rate. That is, we have a monthly interest rate of $(1/12) \times 0.06 = 0.005$, or, 0.5% per month. Now over the long term it makes a big difference whether the interest is *simple*, which means that you never earn interest on the interest, or whether it is *compounded*, where the interest is put back into the account, and you earn interest on the total amount, including all previous interest payments.

For simple interest, the amount is 0.5% of the principal, $12,000, and again remembering that "of" means multiplication, I earn $0.005 \times \$12,000 = \$60$ interest each month, which I am free to take out and use as I please. So over a year I earn $12 \times \$60 = \$720$ interest. Not bad. What about if the interest payments are compounded? The interest payment for the first month is the same as the simple interest, $0.005 \times \$12,000 = \$60$. But the next month I earn interest on both the \$12,000 principal and the \$60 interest from the previous month, that is, on \$12,060. The interest, still at 0.5%, on \$12,060 is $0.005 \times \$12,060 = \$60.30$. Wow – an extra 30 cents by talking compound interest over simple interest. Whoopee.

The next month I earn interest on everything so far, $\$12,060 + \$60.30 = \$12,120.30$. The interest is $0.005 \times \$12,120.30 = \$60.6015$, an extra 60.15 cents over the simple interest. I could go on and calculate the interest month by month for the whole year, but why do a lot of grunt work that mathematics can save me from? The total each month is the previous balance, which we'll call b, plus the interest earned, which is 0.5% of the previous balance, that is, $0.05b$. So the total in the account is, after the interest payment, $b + 0.005 \times b$. Now back in school, we all learned that common factors can be pulled out, and here b is a common factor, if we write the first term as $1 \times b$. So the total amount in the account after the interest payment is $1 \times b + 0.005 \times b = 1.005b$. That means, the total amount after the next interest payment is 1.005 times the previous amount in the account. This is really handy.

Let's check our formula. After the first interest payment on the original \$12,000 balance we had \$12,060 in the account, and $1.005 \times \$1,000 = \$12,060$. After the second payment we found that we had \$12,120.30, and $1.005 \times \$12,060$ is indeed \$12,120.30. Good! I get a certain amount of pleasure from having my mathematics agree with the facts. If I start with \$12,000, after one month I will have $1.005 \times \$12,000$ in the account. The next month I will have 1.005 times this amount, which is 1.005×(1.005×$12,000). Rather than replace 1.005×$12,000 by $12,060, I'll keep it as is, as I can see that this amount can be written as 1.005×1.005×$12,000 (I don't need brackets when only multiplication is present) and using exponents, this is 1.005^2×$12,000. At the end of the third month will be 1.005 times this amount, or 1.005^3×$12,000. The pattern is evident now; after n months, the amount in the account will be 1.005^n×$12,000, so after one year, or 12 months, the amount in the account will be 1.005^{12}×$12,000. Using a calculator, I can see that this turns out to be approximately $12,740.13. The advantage of using exponents is that you don't need to multiply 12,000 by 1.005 twelve times; you just multiply 12,000 by 1.005 raised to the twelfth power.

I like to go through the steps rather than memorize the formula for compound interest, as I like to keep my mind as uncluttered of formulas as possible. A brain cluttered by formulas is *not* a beautiful mind. There is so much more to learn by doing the deduction, and some pleasure in using my mind to find the formula.

LEND ME YOUR EAR

Lending works on the same principles, except it's a bit more complicated. Have you ever checked a loan or car payment schedule to find out how they arrive at the monthly payments and the buyout? Very few people do. You have trust. But should you?

I believe in the principle "trust but verify." Let me take you through the details of a recent five-year loan I took out, with monthly payments. The principal amount I borrowed was

$12,150.20 at an annual yearly rate of 20.99%. Wow, that is high, but if the loan is unsecured, you are probably looking at a rate at least that high. I find it funny that the annual rate is 20.99% rather than 21%; do financial institutions fool a lot of people into thinking the rate is lower than it is?

How can I calculate what the monthly payments should be? The contract tells me, but remember my motto – "trust but verify." It's not so simple as calculating the monthly interest, as I wouldn't be paying off just interest each month – I would pay off a bit of the principal as well. First of all, the loan was taken out on January 12 of a non-leap year, with the first payment (of 60) due on February 27. The way to look at this is essentially to have interest carried over for 15 days until January 27, and then start the loan. The interest due on January 27 is 15 days of a rate of 20.99%. But this rate is for a whole year, so on a daily basis, it is a rate of $(1/365) \times 0.2099$, which is about 0.000575, or 0.0575%. So the interest due on January 27 is $15 \times 0.000575 \times \$12,150.20 = \$104.80$. Now on that date I have 60 equal monthly payments I have to make, over five years.

At this point I don't mind looking up or remembering the formula for the monthly payments, but again this is something I can derive. First of all I need a formula for adding up a *geometric* series $(a+ax+ax^2+ax^3+ \ldots +ax^n)$, which is a bunch of numbers where the first term is some number a, and to get the next number from the previous one, you multiply by the same fixed amount, x, called the *common ratio*. Either you memorize this, or even better, you memorize the "trick" that we used in our Achilles-tortoise example of Chapter 8. If we call this sum S, then we multiply it by the common ratio and subtract, we get the following:

$$S \quad = \quad a + ax + ax^2 + ax^3 + \ldots + ax^n$$
$$xS \quad = \quad \phantom{a + {}} ax + ax^2 + ax^3 + \ldots + ax^n + ax^{n+1}$$
$$\overline{(1-x)S \quad = \quad a -ax^{n+1}}$$

I made sure that when I multiplied all the terms in the sum, I moved them over so that I could see what cancelled out when I subtracted. On the left side I have $S-xS$, which I rewrite as $1S-xS = (1-x)S$. The right side I can rewrite in a similar manner as $a(1-x^{n+1})$, so by dividing both sides through by $1-x$ and replacing S by the series it represents, I get the formula for adding up the first n terms of a geometric series:

$$a + ax + ax^2 + ax^3 + \ldots + ax^n = a\frac{(1-x^{n+1})}{1-x} = a\frac{(x^{n+1}-1)}{x-1}.$$

To tell the truth, I have used this formula so many times that I know it well by heart. But the trick is so nice, and so useful, that I couldn't help but show you.

Back to the loan payments. I want to make the same monthly payment so that the loan is paid out after five years, which is 60 payments (12 payments per year). The way to think of it is, if each of the payments were put into an interest-bearing account at a rate of 20.99% per year, the total after the 60 payments in the account should be the principal, $12,150.20, plus the interest accrued over the 60 payments. Since the monthly interest rate is $0.2099/12 = 0.017491667$, or about 1.749% per month, the amount due after the 60 payments is $(1+0.017491667)^{60} \times \$12{,}150.20 = \$34{,}390.23$. This is why it is so important to pay off some principal as you go along. Otherwise you will have a whopping bill to pay at the end.

Now on the other hand, suppose our equal monthly payments are called K (this is exactly the amount we want to find). The first payment, paid at the end of the first month, would accrue 59 months of interest until the end of the loan, and so, at a monthly interest rate $r=0.2099/12$ (we will use r for now instead of the number just to keep things tidy), it would amount to $K(1+r)^{59}$ if invested at the same interest rate (compounded monthly). The second payment would likewise amount to $K(1+r)^{58}$, and so on, until the final payment, K, which would accrue no interest. The sum total of these amounts should equal the total amount that the loan accrues to, so we get an equation:

$$K(1+r)^{59}+K(1+r)^{58}+ \ldots +K(1+r)+K=34{,}390.23.$$

The left side is just one of those geometric series we talked about, written in reverse order, with first term K, common ratio $1+r$ and $n=59$. By the formula we found earlier, it sums to

$$K\frac{(1+r)^{60}-1}{(1+r)-1}=K\frac{(1+r)^{60}-1}{r} \text{. So the equation takes the form of}$$

$$K\frac{(1+r)^{60}-1}{r}=34{,}390.23 \text{, and a little bit of algebra finds } K\text{:}$$

$$K=34{,}390.23\frac{r}{(1+r)^{60}-1} \text{.}$$

To finish off, all we need to do is plug in the value of r (which is $0.2099/12=0.017491667$), and we get $K=328.64$. What does the contract show? The first monthly payment, on February 27, is \$433.43 (which is the interest of \$104.80 for the first 15 days plus the first monthly payment of \$328.64) plus 59 monthly payments of $328.63. Both of these numbers agree with my calculations to within a penny, and I can't expect much more accuracy than that.

The amount of interest due if I don't make a payment for five years is quite astounding: about $22,000 on my $12,000 loan. That's why I'm always leery about those furniture and electronic companies' "Don't Pay Until . . . Events" that are advertised, where you don't need to make a payment for a couple of years or so. Read the fine print. You'll probably find that while you don't pay any interest provided you *pay the total amount off on its due date*, if you can't or don't pay it all out at that time, you are responsible not just for the interest from that point onwards, but for all interest accumulated since you signed the deal. And as we've seen, that can add up to a bundle, much more than you ever imagined. The furniture companies will have you by the proverbial chandeliers.

WHAT'S IT WORTH TO YOU?

Loans prove a point: money and time depend on one another. One hundred dollars in your hand now isn't (and shouldn't) be worth the same as $100 five years ago, or $100 five years from now. Suppose I had $100 five years ago, and the interest rate has been 10% consistently since then. This is a simplification, as interest rates fluctuate from day to day, but I don't want the complexity to get in the way of the point.

We mathematicians do such simplifications all the time. I remember a joke that goes something like this. A physicist, a statistician, and a mathematician were at the racetrack, each trying to figure out which horse to place a bet on. The physicist was busy finding out all of the relevant variables – wind speed, weather, track conditions, the various forces acting on each horse, and so on. The statistician was looking over reams of data on previous horse races. Suddenly the mathematician shouts, "I've solved it! I know which horse to bet on." The others gather round, and the mathematician starts, "First, let's assume each horse is a perfectly spherical object . . ."

But I digress. Back to the worth of money. If the interest rate has and will be 10% per year over the foreseeable future, then $100 five years ago should be worth $1.10^5 \times \$100 = \$161.05$ today, as that is what it would have been worth had you just banked it at 10% (compounded yearly). So what about the present worth of money in the future? For example, what is the present worth of $100 five years from now? Well, if it is worth x dollars now (when faced with an unknown, at least give it a name for heaven's sake), then in five years, if invested at 10% interest compounded yearly it will be worth $1.10^5 \times \$x$. But this is supposed to be $100 in five years, so we get the equation $1.10^5 \times \$x = \$100$, or $\$x = \$100/1.10^5 = \$62.09$, that is, $100 in the future is worth only $62.09 dollars now.

It may sound esoteric, but companies use present values of money to weigh their options, and I think that they are something that everyone should be aware of. Otherwise, when you need to evaluate your options for money now or money later, you may be at a disadvantage.

MONEY WORRIES

Sometimes there are bigger money worries than how much my money is worth. I do think about the security of all of the transactions I carry out, whether at local stores or over the Internet.

On the one hand, I have become so reliant on the ease of Internet banking that I am loath to give it up, but on the other hand I realize that I am at a big disadvantage not knowing who might intercept various wireless signals between myself and the bank. I still carefully consider whether to give out my credit card information over the Web, as it is a bit of a black box out there; I do not know who has the information or where it is stored. There is a booming market in scam artists posing as legitimate companies, even banks, asking you to log on to their secured web page, which has a URL very similar to the one used by the real institution, and to change your password or enter your personal information. Of course, the information is collected and used for fraud.

While I am wary of that type of scam, the general problem of the security of information transmitted electronically interests me, on a number of levels. I am happy to trust it, provided I feel it is trustworthy. To keep prying eyes from messages, data needs to be encrypted, which means that it needs to be encoded in such a way that it is difficult, if not impossible, for anyone to decode (except those who should be able to decode it). Encryption has a long history; there has always been a need to hide information from someone's eyes. In ancient Greece, war plans were encrypted by wrapping papyrus around a cylinder and writing the code downwards; the unfurled sheet could only be read by rewrapping it around another cylinder of the same diameter. The code breaking of German messages by Allied cryptologists hastened the end of the Second World War, by many accounts.

The principles of encryption, that is, the encoding of information, are fairly elementary. You need some procedure, some algorithm, for transforming any message into the encrypted message that will be transmitted. The intended receiver must be able (and usually, in fairly quick fashion) to decode the message, while anyone else intercepting the message should be unable to decode it.

Here's an easy starting point. You may even have tried encoding messages using a similar technique when you were passing notes back and forth in class in elementary school and junior high. What you do is list the alphabet in a row, along with some punctuation, say a space, comma, colon, period, and question mark; now you have 26+5=31 symbols in total that you can use in any message.

A B C D E F G H I J K L M N O P Q R S T U V W X Y Z _ , : . ?

Then you and your friends all agree on a number between 1 and 30, say 7, which we call the *key*. For any message, take each symbol in turn, and replace it by the symbol that is moved the key number of spots to the right – in this case seven spots – wrapping back to the beginning when you reach the last symbol.

So, for example, suppose the message is "CALL ME TONIGHT." You would encode C as J, as the key is 7 and the symbol seven positions to the right of C is J. Similarly, A is encoded as H, and so on. The space, which appears in the list as _, is encoded as C, because to move seven symbols to the right of the space in the list, you move four places to the right, to the end of the list, and then three more from the beginning of the list to get to C. The entire message sent is "JHSSCTLC_VUPNO_F." That's pretty cryptic!

The decryption procedure is just as easy; move the corresponding symbols back seven places in the list, wrapping around from the beginning of the list to the end if necessary. J, the first letter in the message, corresponds to C, and so on.

You can make a decoder – this is very popular with kids – by taking two circles of paper, one slightly larger than the other, and putting a tack through the centres. Both wheels should have the symbols listed clockwise around the outside of the wheel. Start by aligning the symbols on each wheel. Then to encode, turn the inside wheel the key number of places counterclockwise and read off the corresponding symbols, one by one, on the inside wheel. To decode, after aligning the symbols again, turn the inside wheel clockwise the key number of places, and read off the corresponding symbols, again on the inside wheel.

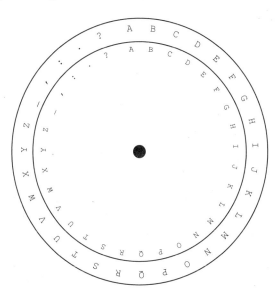

ENCRYPTION WHEEL

What I have described here is a pretty easy encryption procedure. The only drawback is that if someone knows that you are using a 31-symbol encryption wheel, then they can probably decode the message fairly easily by trying, in turn, all possible keys (from 1 to 30) to decode the message. Probably only one of the 30 decoded messages will make sense, and that one is the original message.

There are even more subtle tricks you can play to decode messages. One relies on the fact that in the English language certain letters appear on average more often than others in sentences. For example, the letter E typically appears more often than any other letter, by a factor of about 50%. Such rules of thumb can greatly reduce the amount of different keys you'd likely need to consider to decode the message.

IN YOUR PRIME

The encryption schemes for Internet transactions ought to be more secure than an encryption wheel, and they are, thank goodness. The details involve an area of mathematics called number theory, which revolves around properties of numbers, and in particular, prime numbers. (Just in case you forgot, a whole number bigger than 1 is prime if the only whole numbers that divide it evenly are itself and 1.) The first few primes are easy to list: 2, 3, 5, 7, 11, 13, 17, 19. It involves a little more work to list all the primes under 100. The ancient Greeks knew how to do this, and the technique is known as the *Sieve of Eratosthenes*. To look for primes up to 100, you must first list all the numbers from which you want to sift out the primes, starting at 2.

	2	3	4	5	6	7	8	9	10
11	12	13	14	15	16	17	18	19	20
21	22	23	24	25	26	27	28	29	30
31	32	33	34	35	36	37	38	39	40
41	42	43	44	45	46	47	48	49	50
51	52	53	54	55	56	57	58	59	60
61	62	63	64	65	66	67	68	69	70
71	72	73	74	75	76	77	78	79	80
81	82	83	84	85	86	87	88	89	90
91	92	93	94	95	96	97	98	99	100

Then you take the first number, 2, and cross out all multiples of that number (4, 6, 8, and so on up to 100) until you reach the end of the list. These are not prime (as 2 divides them), but 2 is.

	2	3	4̶	5	6̶	7	8̶	9	1̶0̶
11	1̶2̶	13	1̶4̶	15	1̶6̶	17	1̶8̶	19	2̶0̶
21	2̶2̶	23	2̶4̶	25	2̶6̶	27	2̶8̶	29	3̶0̶
31	3̶2̶	33	3̶4̶	35	3̶6̶	37	3̶8̶	39	4̶0̶
41	4̶2̶	43	4̶4̶	45	4̶6̶	47	4̶8̶	49	5̶0̶
51	5̶2̶	53	5̶4̶	55	5̶6̶	57	5̶8̶	59	6̶0̶
61	6̶2̶	63	6̶4̶	65	6̶6̶	67	6̶8̶	69	7̶0̶
71	7̶2̶	73	7̶4̶	75	7̶6̶	77	7̶8̶	79	8̶0̶
81	8̶2̶	83	8̶4̶	85	8̶6̶	87	8̶8̶	89	9̶0̶
91	9̶2̶	93	9̶4̶	95	9̶6̶	97	9̶8̶	99	1̶0̶0̶

Move on to the remaining numbers, following the same steps. Take 3 and cross out all multiples of 3: 9, 15, and so on (some multiples, like 6, have already been crossed out).

2 3 4 5 6 7 8 9 10
11 12 13 14 15 16 17 18 19 20
21 22 23 24 25 26 27 28 29 30
31 32 33 34 35 36 37 38 39 40
41 42 43 44 45 46 47 48 49 50
51 52 53 54 55 56 57 58 59 60
61 62 63 64 65 66 67 68 69 70
71 72 73 74 75 76 77 78 79 80
81 82 83 84 85 86 87 88 89 90
91 92 93 94 95 96 97 98 99 100

After a bit of work, you will end up with the following:

2 3 4 5 6 7 8 9 10
11 12 13 14 15 16 17 18 19 20
21 22 23 24 25 26 27 28 29 30
31 32 33 34 35 36 37 38 39 40
41 42 43 44 45 46 47 48 49 50
51 52 53 54 55 56 57 58 59 60
61 62 63 64 65 66 67 68 69 70
71 72 73 74 75 76 77 78 79 80
81 82 83 84 85 86 87 88 89 90
91 92 93 94 95 96 97 98 99 100

The numbers that remain after you have sifted out the multiples are the primes (hence the name Sieve of Eratosthenes): 2, 3, 5, 7, 11, 13, 17, 19, 23, 29, 31, 37, 41, 43, 47, 53, 59, 61, 67, 71, 73, 79, 83, 89, and 97.

The study of prime numbers has always been one of the

most fascinating areas of mathematics. Prime numbers are easy to describe but defy any formula. There is just something mystifying about them; they are at once completely determined yet also somewhat random in that there seems to be no way to easily predict the next prime.

Even with computers, it is not easy to find the next prime. Of course, we know fairly large primes now, up to about 1 million digits. A well-known proof shows that there are infinitely many primes. Suppose there were only finitely many primes, which we could therefore list, as say $p_1, p_2, \ldots, p_n$. Then consider the number we get by multiplying all of these numbers and adding 1 to the result, $p_1 \times p_2 \times \ldots \times p_n + 1$, which we'll call N. This number, being bigger than all of the numbers in the list of primes, couldn't be prime. But if a number isn't prime, then a prime number divides it, so one prime from our complete list of primes – $p_1, p_2, \ldots, p_n$ – must divide N evenly. On the other hand, none of the numbers $p_1, p_2, \ldots, p_n$ divides $N = p_1 \times p_2 \times \ldots \times p_n + 1$ evenly, as the remainder is always 1. So we have a big problem, a contradiction. Something is wrong, and the only possible thing that could be wrong is our assumption that there are only finitely many primes. Therefore there must be infinitely many primes!

This is an argument by contradiction, and doesn't really help us find larger and larger primes; it only tells us that the list never ends. In fact, it's rather hopeless to try to find a simple function like a polynomial that will spit out only primes; it can't be done. It is known *about* how many primes there are less than a number, and *about* where the primes are located, but exact values are difficult to come by for very large primes.

Now what does this have to do with encrypting? Well, banking transactions are encrypted with a procedure known as RSA, after its discoverers, Ron Rivest, Adi Shamir, and Leonard Adelman. It is based on the fact that factoring a whole number (that is, breaking down a number into a product of primes) seems to be an intrinsically difficult task. Small numbers are pretty easy to factor: 91 is 7 times 13. But large numbers pose more of a problem, even if you know that they have just two factors. Suppose I tell you that 22,935,755,729 has two factors. Does that help you factor it? By the state of mathematics so far, the answer is no. In fact, it would probably take a fair bit of time, even with a computer. How did I produce the number? Well, I took two large primes, which happened to be 104,729 and 219,001 and multiplied them. But without knowing the primes I used, it would be very time consuming to factor the number.

Large prime numbers are an essential part of how the encoding of Internet transactions works. Even if an Internet transaction is intercepted, it would take much more than a lifetime, and a huge amount of computing power, to decode it. That is why finding larger and larger primes is so important, because the larger the primes used, the longer it should take anyone to factor the number and hence break the code.

How comfortable does the whole RSA encryption process make me feel about my Internet transactions? Pretty darned secure, as I am confident that factoring numbers has, over many centuries, attracted the best mathematical minds, without a solution. But that doesn't mean that some math genius won't come along with some weird, brilliant, new way to factor numbers that could make all Internet transactions public information. It's just that I think it's unlikely. I think I'll worry first about being hit by a meteor.

SAFETY IN NUMBERS

While I'm thinking about security, I am sitting at my computer trying to log in to yet another site that I have visited before. Yet again, I can't remember whatever password I used when I first entered the site. Damn! Why do all these sites require passwords? I can see why I need them at sites where I purchase things, but for entering contests? Patiently, I go through my usual list of passwords, hoping one will do the trick.

Criminals often have an easy time guessing passwords for accounts. Sometimes it is just the default password that has been set up, like the word "system," or a name of the installer. Hackers use computer programs to try a succession of passwords, running through the entries in dictionaries of names and words, because most people choose passwords that mean something to them.

For my really important passwords, I can use randomness to make them as secure as possible. Let me explain. There are, I think, approximately a quarter of a million English words, including most proper names. That's a big number, but not too big for computers nowadays. But if we look at strings of, say, eight letters, perhaps just lowercase letters, we get almost 209 *billion* such strings, and with 10 letters, the number is about 141 *trillion*, an enormous number. So if I need a really secure password, I just choose a random string of 10 letters, something like "gajjueyxmn." The chances of anyone, even a computer, guessing it correctly on the first try are astronomically small. The chances that your random string is a word in the English language is also tiny, so criminals won't find it that way either. There is safety in randomness.

TRUST IN NUMBERS

Accountants need to trust numbers, don't they? People submit all sorts of expenses claims and income tax forms, and accountants are expected to check everything. Accountants can check the calculations themselves and that the forms are filled in properly, but is there a way to check the reasonableness of the numbers entered? That would certainly seem to be a sticky problem. Numbers can come from anywhere, and who is to say that one number should be another? Aren't expenses and other financial numbers random to a large extent?

The surprising answer is no, they aren't quite random. The story began a long time ago when Nova Scotian mathematician Simon Newcomb noticed at the library that the book of logarithmic tables was not evenly worn; the pages nearest the beginning were dirtier than the rest. Before the advent of calculators, books of logarithmic tables were used to help carry out calculations. Logarithms are mysterious, people-loving mathematical pets. They bring large numbers down to reasonable ones. For instance, the number of digits you need to write down a positive integer is the number you get when you round the "logarithm base 10" of the number up (if it happens to be an integer, you push it up to the next integer as well). So, for example, you need six digits to write 946,723, and my calculator tells me that the logarithm base 10 of 946,723 (written $\log_{10} 946{,}723$) is about 5.976222928. In addition to bringing large numbers down in size, logarithms simplify arithmetic – they convert multiplications into additions, divisions into subtractions, and powers into multiplications. Back before calculators, you couldn't call yourself a scientist without having an intimate knowledge of logarithms.

But this leaves the mystery of the book of logarithmic tables. The pages at the beginning of the book, for those numbers starting with a 1, were the most worn, while the pages at the end of the book, starting with a 9, were the least worn. What accounted for this?

It appears that numbers don't appear equally often in many situations. For example, for accounts of money that grow from year to year due to interest, it is much more likely that the first number will be 1 than 9. For example, if the interest rate is 5% per year, and you start with $982, then the bank account will have the following closing balances over the first 50 years (reading down each column):

982.00	1,851.71	3,491.67	6,584.07
1,031.10	1,944.29	3,666.25	6,913.27
1,082.66	2,041.51	3,849.57	7,258.93
1,136.79	2,143.58	4,042.05	7,621.88
1,193.63	2,250.76	4,244.15	8,002.97
1,253.31	2,363.30	4,456.35	8,403.12
1,315.97	2,481.47	4,679.17	8,823.28
1,381.77	2,605.54	4,913.13	9,264.44
1,450.86	2,735.82	5,158.79	9,727.66
1,523.40	2,872.61	5,416.73	10,214.05
1,599.57	3,016.24	5,687.56	10,724.75
1,679.55	3,167.05	5,971.94	11,260.99
1,763.53	3,325.40	6,270.54	

You can see that most of the numbers begin with the digit 1, and fewest begin with the digit 9. The following bar graph summarizes the relative frequencies of the first digits:

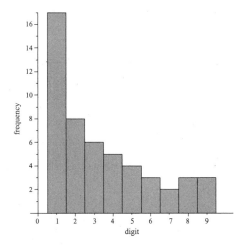

These are obviously *not* random. *Benford's Law* (named after physicist Frank Benford) states that the distribution of first digits in sets of numbers that often arise in accounting settings follows a certain pattern:

FIRST DIGIT	PROPORTION
1	30.1%
2	17.6%
3	12.5%
4	9.7%
5	7.9%
6	6.7%
7	5.8%
8	5.1%
9	4.6%

This pattern has shown up so often that auditors use it to get an idea of whether the numbers submitted to them have been "cooked." Auditors don't expect numbers to correspond perfectly to Benford's Law, but the larger the data set, the more closely it should follow the rule. In fact, data sets as diverse as stock prices, city populations, and bank account balances often follow Benford's Law quite closely. Of course, there are others that won't, such as people's height in feet – there isn't a predominance of people whose height in feet begins with the digit 1, and no one's height begins with a 9! For Benford's Law to hold, there can be no limits (natural or unnatural) on the values of numbers.

Just in case someone shifty might be wise to Benford's Law, there are higher-order principles that state proportions about the first two digits of numbers, and so on. On the other hand, for most of these data sets, the proportions of last digits, and pairs of last digits, behave rather randomly, unlike the first digits. And this gives auditors yet more ammunition. They might scan through a large number of expense claims for a company, noting that among the last digits, the proportion of numbers ending in 00 might be much more than expected (you would expect roughly only 1% of the numbers to end in 00). This might be a sign that employees are cooking some of the numbers, or rounding.

Or a computer program might find a predominance of 4s in the hundreds digit. Perhaps this has something to do with the fact that, at this company, expenses under $500 don't require an executive signature. Then again, it might mean nothing, and be just due to chance variation. It's up to the auditors whether to investigate. In any event, with all of these laws, it is foolhardy to try to beat the system. Humans just can't cook up numbers that match real-life data. People who try to invent "random" numbers inevitably have biases, liking some digits more than others. In the end, mathematics will out them. So in numbers we can indeed trust.

I've had enough of paying bills and counting shekels for today. All this thinking about Internet security has made me a little jittery, and I think I'll go visit one of the online sites, http://primes.utm.edu, that keep track of the latest on all things prime. It seems that with yet more work and computer time, someone has found another large prime, this time with close to 13 million digits. The bigger they are, the harder they fall.

NATURE, ART, AND FRACTALS

Dinner's over, and I am out taking a walk around my neighbourhoood. Even though I live in a fairly big city, at least by Canadian standards, I have a forest and a pond practically in my backyard. Many times I have seen deer come to the edge of the forest and wait to cross traffic. Several ducks waddle up to me, looking for a handout, and a few aggressively come after me. I think the biggest one is chasing me with a squeegee.

I love looking at trees. Their beauty really captivates me. I used to take my sketchbook and sit for hours drawing, but I seldom have time for that now. I used to think that mathematics and art were diametrically opposed, but I am finding more and more how closely related they are; I am coming back to art via mathematics. Sometimes life imitates art, but now I am finding that math imitates art (and quite well, I might add).

Almost everyone thinks of mathematics as a science. You know – math is left brain, art is right brain. But I consider myself more of an artist than a scientist. Now that I think back, my childhood interest and ability in art, and in drawing in particular, followed a parallel course to my interest in mathematics. What attracted me to both was aesthetics, not utility. Mathematics had the added benefit of being useful, but that wasn't its chief attraction.

Both art and mathematics mix right and left brains, artistic creativity with logical and scientific reasoning. There are rules and principles artists must adhere to for things like perspective and colour, while for mathematicians the very best arguments are wildly creative and connect ideas in unexpected ways.

I find it frustrating how little the public appreciates mathematical beauty. Most people think of mathematicians as being only slightly more interesting than accountants. In the movie *Stranger than Fiction*, Will Ferrell's character, an auditor, laments the humiliation of his girlfriend having left him for an actuary.

I knew from quite early on that my mathematical ability was not something to advertise around women. Luckily, I had an artistic and musical side.

THE BEAUTY OF MATHEMATICS

Before the 1900s, a significant segment of the population appreciated mathematics as an art. The Greeks had mathematics as one of their five main fields of study, and a knowledge of mathematics was considered a sign of a well-bred person. Some of the best math researchers used to be people outside the profession – lawyers, accountants, priests, and so on. People used to do mathematics, dare I say it, for the *fun* of it!

I know a researcher, Robert, a theoretical physicist. He is your typical brilliant scientist – a wiry, skittish man who thinks and converses at warp speed. The last talk I heard him give went over the time limit by about 20 minutes, which is a big no-no. Goodwill diminishes exponentially with every minute you speak over your allotted time. But I, like everyone else, was enjoying the show. Robert was showing slide after slide in rapid-fire succession, stating over and over again, "And for my last slide . . ." His enthusiasm for his mathematical research was infectious – you couldn't help but be carried away.

THE MATHEMATICS OF ART

When I look at scenery, whether in artwork or on a walk, I often see some underlying mathematics. We have a trail called the Power Line in our neighbourhood that seems to stretch forever, and I can't help but notice how the trees lining the path seem to converge off in the distance. Perspective is geometric at heart.

When deciding on the composition of a three-dimensional drawing or painting, artists choose a point, called the vanishing point, that lies on the horizon. It is the place at which all parallel lines that move from the foreground into the background meet.

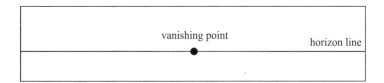

Now that might seem wrong, as you know from all those geometry classes you had that parallel lines *never* meet. But in perspective drawings they *look* as if they meet, as they appear to do in real life. If you look at a pair of straight train tracks that lead away from you (note: please stand off the tracks!), they will *appear* to meet at some point off in the distance. At what point do they meet? At the point of infinity. But rather than bend your brain anymore, we'll just accept that the parallel lines appear to meet in the distance at this vanishing point. The artist might add a couple of railway tracks moving off into the distance, meeting at the vanishing point, and a few railway ties, perpendicular to the tracks, and parallel to the horizon. They appear as parallel horizontal lines, but they get smaller as they get farther away. This is done by making each successive tie, say, 80% the size of the previous one, centring them on the tracks, and making sure they appear closer together as they get farther away.

Vertical objects are vertical in the drawing, but lines that move from foreground to background should be extended to meet at the vanishing point. I drew in a couple of guide lines for the tops and bottom of the telephone poles, which I'll remove in a moment, and then placed the poles appropriately on them.

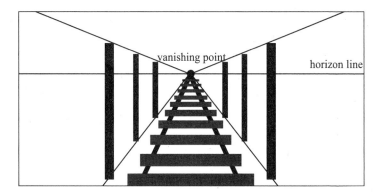

Now I can add in some wires on the poles, in the same way.

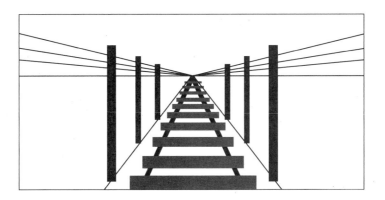

I think you get the idea. The geometric principle of perspective drawing is much more than theory: it can help anyone make a picture that looks three-dimensional.

This reminds of a time when my younger brother and sister were playing tetherball in the backyard. My sister slammed the ball, and it hit my brother square in the face. He grabbed his nose with both hands as blood poured down his arms and dripped from both elbows. My sister, distraught, ran into the house, and screamed, "*Why does everything always happen to me?*" Perspective really is everything.

THE GOLDEN RULE

Another geometric principle used in creating both sculptures and paintings has to do with what is considered the perfect proportions of objects. The ancient Greeks felt that a certain proportion for a rectangle was the most aesthetically pleasing:

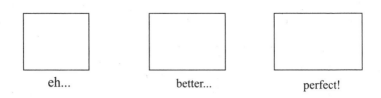

eh...   better...   perfect!

The ratio of the larger side to the smallest in this perfect rectangle is approximately 1.618, and is called the *golden mean*, the "golden" part referring to its apparent perfection. The number actually is *irrational*, which doesn't mean it's argumentative, but rather that it isn't a fraction of whole numbers.

The golden mean is often referred to in mathematical circles as *phi*, or by the Greek symbol ϕ. There are lots of ways to define it precisely. Here are a couple:

1. Divide a line segment up into two segments, with the property that the ratio of the length of the whole line to the larger segment is equal to the ratio of the larger to the smaller:

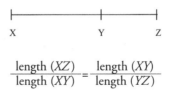

$$\frac{\text{length } (XZ)}{\text{length } (XY)} = \frac{\text{length } (XY)}{\text{length } (YZ)}$$

So if we let *YZ* have length 1 and *XY* have length ϕ, then *XZ* has length $\phi+1$ and so the equation reads $\frac{\phi+1}{\phi} = \frac{\phi}{1}$. A little bit of algebra tells us that this is the same as $\phi^2 - \phi - 1 = 0$.

2. Suppose you want to create a rectangle for which, if you break it up into a square and a smaller rectangle, the

smaller rectangle will have the same proportions as the bigger rectangle (you can repeat this process forever, if you like).

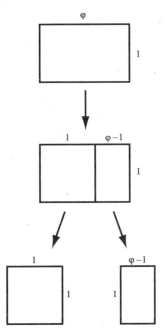

Here we want the smaller rectangle, whose larger side is length 1, and whose smaller side is length $\phi-1$, to be in the same proportion as the original rectangle, whose larger side is length ϕ and whose smaller side is length 1: $\frac{1}{\phi-1} = \frac{\phi}{1}$. After a bit of algebra, we get the same equation we got in the first case; namely, $\phi^2-\phi-1=0$.

Now to find the value of the golden mean, all I need to do is solve the equation $\phi^2-\phi-1=0$. It is of the type you must have seen before, a quadratic. You probably remember learning something like the "quadratic formula," which was a rule to solve exactly such equations. If you apply the rule you get that $\phi = \frac{1+\sqrt{5}}{2}$ (actually, there are two answers, but the one we are after is the number that is bigger than 1). The square root of 5 is what makes this irrational. Using a calculator, you find that ϕ is about 1.618, as I mentioned earlier.

The golden mean was so beloved that artists used its proportions in many, many works of art in a variety of ways. Human proportions were often set by the golden mean. It has been observed that various proportions in Leonardo da Vinci's paintings (including the pretty face of the *Mona Lisa*) are in the golden ratio. Architects, in designing buildings, also used the idealized proportions in their blueprints. The golden ratio appears in various guises throughout the Parthenon.

The golden mean has a number of interesting properties. First of all, the reciprocal of the golden mean is 0.618 . . . , what you get when you subtract 1 from it. Another property involves a famous sequence of numbers, called the *Fibonacci sequence*: 1, 1, 2, 3, 5, 8, 13, 21, ... You start off with two 1s and each new term is the sum of the previous two terms. You have likely seen the sequence before, or at least heard about it. If you divide one number in the sequence by the previous one, you get 1, 2, 1.5, 1.67, 1.6, 1.625, and so on, with the numbers getting closer and closer to, you guessed it, the golden mean.

Musicians, too, may have found inspiration in the golden mean. Composers Béla Bartók and Claude Debussy seem to have incorporated the golden mean into their music, either consciously or subconsciously. But more about music in Chapter 12.

WHAT IS BEAUTY?

Whether a rectangle is or isn't attractive inevitably gives rise to the question What is beautiful? There probably isn't one answer. But I think that whether you are mathematically inclined or not, there are certain things that most people find pleasing.

One is symmetry. If you look at someone's face or body, you expect that the left side will be approximately the same as the right side. Picasso's distorted faces may have artistic appeal, but I don't think many men would want to date a woman with both eyes on the same side of her nose. Apparently symmetry is sexy as well. A study has found that people whose faces and bodies are more symmetrical tend to start having sex earlier and have more partners. On the other hand, beauty is still in the eye of the beholder. Lyle Lovett's face is not symmetrical, and he hooked Julia Roberts. So go figure.

The mathematical notion of symmetry involves movement (or *transformation*) of all the points of an object so that it is carried exactly back onto itself, leaving no part uncovered. The movement also has to preserve the space between points – no shrinkage allowed.

Reflections like the left and right swapping in our faces and bodies are an example of symmetry, but there are other examples in nature, like a rotation by a certain number of degrees. Starfish have this kind of symmetry; you can rotate the starfish about its centre by 72 degrees (or one-fifth of a full 360 degrees) and cover the starfish again.

Symmetry about the origin is another one I like, which is a rotation of 180 degrees about an object's centre, something like the following:

Soon after becoming a mathematician, I became interested in M.C. Escher's artwork, which depicts mathematically based "impossible realities," and which I found to be an inspiring mix of mathematics and art. Symmetry is not the be all and end all of what I find beautiful in art. Most people would think that mathematicians would favour depictions of lines and circles, but these don't have the aesthetic appeal of landscapes for me. Inspiration hit Newton as he was sitting under a tree, not under a polygon.

FRACTALS, FRACTALS EVERYWHERE

If we mathematicians are awed by the beauty and importance of mathematics, we are awed no less by the grandeur of nature. Trees, clouds, and mountains may seem far removed from the building blocks of mathematics – spheres, cylinders, pyramids, and the like – but nothing could be further from the truth. There is a relatively new branch of mathematics called *fractals* that includes many of the beautiful objects we see in nature. Fractals demonstrate a property called *self-similarity*, that is, small parts of the object look just like the whole, only smaller. Have you ever looked at broccoli? I mean, really *looked* at broccoli? Go to a supermarket, pick up a bunch, and pluck off a stalk. It looks very much like the whole bunch. If you take another piece off the small stalk, it again looks like the whole. (You might want to buy the broccoli rather than putting it back.)

Broccoli, trees, clouds, mountain ranges, coastlines – they all have this self-similar property. Pieces look like the whole, and without any background to provide a sense of proportion, you'd be hard-pressed to tell whether you are looking at the entire object, or just a part. Such objects were of keen mathematical interest at the turn of the last century, when mathematicians were looking for what are sometimes called *badly behaved functions*. These functions are naughty in the sense that they appear inexplicable or counterintuitive: a curve that fills an area, or one that is crinkly everywhere but has no breaks in it. And fractals were exactly the creatures they were looking for.

I've found fractals in all sorts of places – shopping in the supermarket, walking outside, looking out the window on an airplane. One of the ways to build fractals is to take a collection of line segments and replace each part by a scaled copy of the whole thing. For instance, suppose we start with the following simple collection of four lines:

After replacing each of the four lines by a smaller version of the whole figure, it looks like the following:

You continue the process, and after a few more iterations, here is what you get:

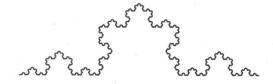

This is a famous fractal, known as the *Koch curve*, after the Swedish mathematician Helge von Koch. The fractal is actually the curve you get after infinitely many steps, but the picture differs little from the one you get after several steps.

This little curve has many beautiful properties, such as being in one piece though it is not smooth anywhere. It also can be seen to have infinite length (the length goes up by a factor of 4/3 every time you make a replacement) though it is contained in a small rectangle.

And the curve does look like a coastline to me, a coastline that would be nearly impossible to draw with standard geometric figures like straight lines and circles. The ability of fractals to model real-life objects has not gone unnoticed by Hollywood. The 1982 movie *Star Trek II: The Wrath of Khan* used fractals to generate various views of the Genesis planet.

Below I used the stick figure at left and the replacement scheme I talked about (replacing every line segment except the vertical one), and generated what I think is a pretty good-looking tree.

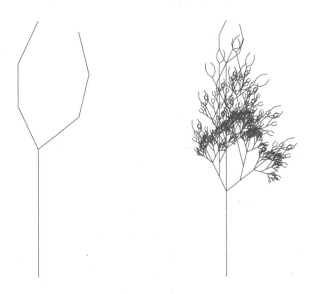

No tree is exactly self-similar of course; there is always some random variation. A fractal-generated tree might have a branch that goes off on a different angle, or have a different length than what the fractal predicts. It seems that we humans like our fractals, but we also like them to have some flaws.

The advantage of having a mathematical description of natural objects is that they can be generated and manipulated by computer, in a way that would be impossible if you used scanned images. The mathematics also suggests a lot more about the ecological and medical advantages of objects. The branching of tubes in our lungs, for example, is fractal-like, providing a large surface area in a relatively small volume (our chests). This branching is

240

crucial to our ability to collect enough oxygen to live. The fractal-like convolutions on the surface of our brains satisfy the same purpose, namely to increase the surface area (where higher levels of thought take place) in a confined space (our skulls). Fractals are not only beautiful, they're essential.

All of this talk about fractals reminds me of a recent mathematical excursion. I had been looking at some research with a previous PhD student of mine, Carl, and my colleague, Richard. It had been one of those "not useful but attractive" problems (what I like to refer to as a "Paris Hilton"). It connected up one part of mathematics, networks, a love of mine, with polynomials, part of both algebra and analysis. While each individual field of mathematics has its own appeal, I find the unexpected connections between them utterly compelling.

I don't know if you recall much about polynomials from high school, but one of the things you can do with them is to find their roots, that is, numbers for which, when you substitute them into the polynomial, you get zero. For example, the polynomial x^2-5x+6 has roots 2 and 3, as $2^2-5(2)+6=0$ and $3^2-5(3)+6=0$. The infamous quadratic formula finds the roots of any quadratic, that is, a polynomial that only goes up to squares in powers.

Anyway, the quadratic formula finds the roots, but sometimes they don't exist. For example, the quadratic x^2+1 has no roots as it is always at least 1, and never zero. But one of the greatest advances in mathematics was to say that there *should* be two roots, but there aren't. So mathematics invented a root, called *i* (for *imaginary*) to allow such polynomials to have roots. The *complex numbers* are formed from all of the real numbers and the imaginary number i. Do these numbers exist? Well, that is a matter of your point of view. In any event, we can treat them as if they exist, and scientists and engineers have found having such numbers around a boon for their research.

Once you introduce complex numbers, all of a sudden all polynomials can be broken down by their roots; it's such an important mathematical fact that it is called the *Fundamental theorem of algebra*. This brings us back to the research problem I was looking at with Carl and Richard.

I was playing around, having the computer find the polynomials, find their roots, and then plot them so I could visualize them. Numbers are fine, but give me a picture any day! I had tried a variety of different things, and then I hit upon an idea. I plugged in my example, and waited for my Mac to show me the results. I had no expectation of whether the picture would be nice or interesting. I simply waited for the surprise.

And what a payoff it was. Here is what filled the screen:

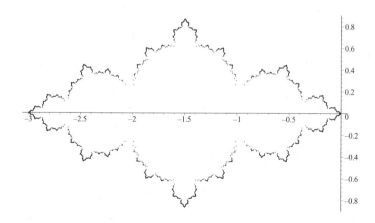

To me that's a beautiful picture – a work of art. It struck me right away that it was a fractal, not what I was expecting at all. I immediately tried out a few more examples, and saw the following:

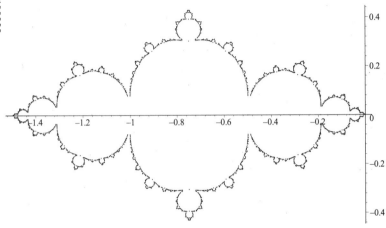

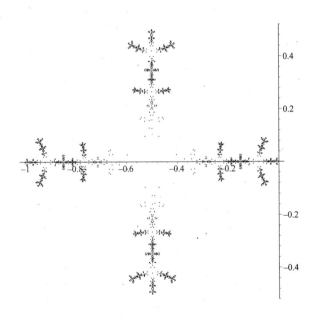

Pretty pictures, and fractals to boot. It took a while for us to find proofs that these objects were indeed fractals, but all of the research began with visualizing mathematics on a computer. Research is indeed an art.

In fact, mathematics without aesthetics is drudgery, endless calculations. When a colleague and I are carrying out research, we fill the blackboard with diagrams and jottings, and wave our hands about for good measure. What a mathematician sees in his diagrams is a lot different from what others see in them. Some research involves breaking down a problem into many, many cases and then working on each case individually. While this laborious approach is sometimes necessary, I always look for a more creative strategy. One of the things I try to instill in my students is love and appreciation for the aesthetics of mathematics. Paul Erdös used to speak of G–d having a book in which all of the most beautiful proofs were inscribed. Occasionally, if we are fortunate, human beings are offered a glimpse into its contents.

I often think about the process of doing research, rather than the research itself. As with any art form, you never know when inspiration will strike, and when it might, in a huff, leave you for another. Creativity comes and goes, and the best you can do is prepare yourself for its eventual return. I find solace in reading the most beautiful mathematics, whether related to my research or not, and I always manage to find some inspiration therein. One of the best compliments anyone can give a mathematician is to declare his research "pretty"; the word "useful" is of lower importance. Being open to what may seem like random or unrelated occurrences is the basis of research; being creative is finding opportunity where others see nothing.

As I make my way back down our street, I see an amazing sight. A house here, a rectangle topped by an isosceles triangle there, a tree here, a fractal there. There is no limit to the beauty that surrounds. Where nature begins and math ends is an interesting question. And sometimes it's wonderful enjoying a question without looking for an answer.

12

SEX, MATH, AND ROCK 'N' ROLL

Friends have called to say that they are in the neighbourhood, and my wife and I invite them over for coffee and dessert. That means tidying up, and for an orderly mathematician, the house is in one state of disorder. I remember a professor I had whose desk was covered by a layer, at least three centimetres thick, of paper. I'd never be like that – or so I thought!

Anyway, with guests coming over, I am always hopeful that I'll be asked to take out my guitar and sing a few songs. I have been a musician and songwriter for almost my entire life. When I was an undergraduate in Calgary, I used to play music six nights a week in bars, getting home at about 2:30 A.M. and being at classes for 9 A.M. Luckily, my first class was calculus, so I could catch up on my sleep.

My love of music is almost as old as my love of mathematics. There was always music in the house. My older brothers played instruments, with one of them being very accomplished on the vibraphone, which sounds more titillating than it is. The vibraphone is a percussion instrument, much like the xylophone, but way cooler, and is used most often by jazz musicians.

I started on piano. I was one of those music students who only practised the half-hour before my lesson each week. My teachers ranged from odd to peculiar. One fell backwards over the piano bench while trying to show me how to be loose at the piano. Another used to feed coins into the fake parking meter my father had placed in the bathroom (one of Dad's many jokes).

When I first heard an Elton John record, I started to pick up sheet music and learn his songs. Then one day, when I was around 12, I heard my first Beatles album and that was it – I forsook the piano and began spending hours teaching myself the guitar, jamming with John, Paul, George, and Ringo. I am self-taught on guitar, having only taken a few lessons back in my teens. (My teacher for those few lessons was a long-haired guitarist whose gaze was fixed, with both eyes staring at his nose. I wonder if it had something to do with the sweet hazy smoke that filled the room where he taught.)

My father was an audiophile, and owned a reel-to-reel tape recorder that had two speeds, one twice as fast as the other. If I copied a Beatles song onto the tape at the high speed and then cut the tape speed in half, the music would play back at half speed, exactly down an octave. It would be years before I would find out the underlying mathematics of this phenomenon, but back then

I was just happy to be able to slow down George Harrison's riffs and solos.

I threw myself into learning the guitar, and within a few years I was playing professionally. Having to get up in front of a crowd and perform was great practice for being a professor. I still enjoy teaching in large lecture halls, as it sometimes feels more like entertaining than teaching.

THE MISSING LINK: MATHEMATICS AND MUSIC

There is a fair bit of research and anecdotal evidence about the correlation between mathematics and music. Children who listen to music early find it easier to understand mathematics. Music is a lot of pattern matching, recognizing themes that are presented forwards, backwards, rotated, and inverted, and mathematics is the study of patterns. It's no coincidence that just as there are child prodigies in music, there are child prodigies in mathematics. Some kids are just hard-wired to absorb and reflect patterns.

Most people erroneously think mathematics is associated with left brain function versus music with right brain function. But mathematics isn't all cold analysis, and music isn't all emotion. If music were all emotion, all we'd see at performances would be some artist blubbering or howling up on stage. Good music relies on clever tricks and structure, and good mathematics is creative and, yes, emotional. I have not come across a mathematician yet who doesn't smile at a pretty proof.

SINE, SINE, EVERYWHERE A SINE

Music is inherently mathematical. All sound is due to vibrations, air molecules moving back and forth in rapid succession. The basic building blocks of all musical sounds are pure tones, which correspond to objects vibrating (and hence the surrounding air molecules vibrating) in a wave.

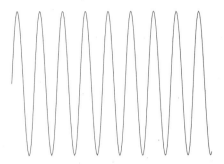

What kind of wave, you ask? In yet another instance of the mathematical nature of the world, it is a *sine curve*. Yes, the same trigonometric function "sine" that you were forced to learn about in high school. If only you were treated back then to the most famous and interesting of all its applications – music. Even if someone knows no mathematics at all, the very act of listening and appreciating music is inherently mathematical.

The loudness of a pure tone corresponds roughly to the sine curve's *amplitude*, that is, how high and low it goes from the midpoint. On the other hand, the pitch of the tone corresponds to the *frequency* of the curve, that is, how fast it repeats. The number of times a sound curve repeats every second is measured in Hertz (Hz). We can hear tones from about 20 Hz to 20,000 Hz, though the range shrinks as we get older.

Now, pure tones are rather boring, much like the sound of a tuning fork. What is so amazing is that all musical tones are made up of such sine curves, formed by adding them up mathematically.

The curves created by adding up a few sine curves form repeating patterns, beautiful to both our eyes and our ears.

It was the ancient Greeks who noticed that tones were made up of many pure tones. Pythagoras and his followers (his "entourage," if he were a rapper) discovered that small fractions were the key to pleasing sounds. Plucking a string that is half as long as the original string produces a tone exactly an octave higher, and perfectly harmonious with the original tone. In fact, it matches so well that men and women often, when trying to sing the same note, sing an octave or two apart. Plucking the string that is two-thirds as long gives another harmony, what we call a perfect fifth, while a string that is three-quarters as long gives us a perfect fourth, and so on. Our ears are naturally inclined to frequencies that are in ratios that are small fractions. We are just built that way. We key in on the ratios of frequencies; if the ratio of two frequencies is, say, 3/2, then they sound like a perfect fifth, regardless of what the frequencies are.

All of these small fraction ratios arise from the *harmonics* to a tone. Typically, every note that an instrument plays is made up not only of a primary or *fundamental* tone, but also of "overtones" that are higher than the original, and these are multiples of the original frequency. If the original frequency is, say, 110 Hz (the frequency of the low open A string on a guitar), then when the open A string on a guitar is plucked, the frequencies 2×110=220 Hz, 3×110 = 330 Hz, 4×110 = 440 Hz, and so on are produced, at different volumes, all combining as the timbre that we associate with the string being plucked. The fact that the harmonics, or overtones, are exact multiples of the original frequency is again one of the wonderful connections between mathematics and music.

Our ears and minds are used to this connection; it is just hard-wired in our circuitry. There is an old trick that church organists employ to play notes off the bottom of the keyboard. Say that I want to play a low 20 Hz tone, and all that is available are the tones 30 Hz, 40 Hz, 50 Hz, and so on. What I can do is strike the two tones for 40 Hz and 60 Hz simultaneously. A listener will hear the two tones simultaneously and will deduce that they are harmonics of a single tone (often called a *combination tone*), which the brain interprets as 20 Hz (the largest number for which 40 and 60 are multiples). This math takes place automatically and subconsciously, even if the listener is math illiterate. Amazing!

HOW IS YOUR TEMPER?

As we have seen, Pythagoras and his crew discovered that the most pleasing-sounding notes, relative to a single tone, correspond to small fractions. But they had a problem. If they repeated the nicest-sounding interval (other than an octave), a perfect fifth, beginning from frequency 110 Hz (a low A), after 12 repetitions, they would get a tone at the frequency $110 \times (3/2)^{12} \approx 14{,}272$ Hz (remember, every time you go up a fifth, you multiply the frequency by 3/2). This sounds suspiciously close to the seventh harmonic of the original A; this harmonic has frequency 110×2^7 Hz. Pretty close, but off enough to be different. The closeness of these two frequencies suggests that they should actually be the same, and that eventually, repeating perfect fifths should bring you back to the original frequency, give or take some octaves. But mathematics dictates that repeating a fifth will never take you exactly to some number of octaves above the original frequency. For if there were some number, k, of fifths that took you to some number of octaves, l, above the original frequency, then you would have to have $110 \times (3/2)^k = 110 \times 2^l$. By cancelling out the 110s and writing $(3/2)^k$ as $3^k/2^k$, it would follow that $3^k/2^k = 2^l$, or by cross-multiplying and adding exponents (which you can do when they have the same base, here 2), we have $3^k = 2^{k+l}$. But this is impossible, as the left side is odd while the right side is even.

Stacking perfect fifths will *never* give you the same result as adding octaves. The difference between the seven fifths and 12 octaves is pretty small, but annoying. It also creates havoc with trying to change keys, as the frequencies of notes depend on how you calculate them. To make a long story short, at some point in the sixteenth century a mathematical solution was proposed. The octave was divided into 12 semitones, with the ratio of each step being the same. If we call the ratio of a semitone r, then going up 12 steps from some frequency f took you to frequency $r^{12} \times f$ (again, going up corresponds to multiplication). But on the other hand, going up an octave takes you to twice the frequency, $2 \times f$. So we need to have $r^{12} \times f = 2 \times f$, which means that $r^{12} = 2$, in other words (drum roll, please) the ratio for a semitone needs to be the *twelfth root* of 2! This number, which is about 1.06, is an irrational number. The scale of 12 semitones, with a semitone corresponding to frequency multiplication by the twelfth root of 2, is called the *equally tempered scale*. It may seem odd to let mathematics dictate music, but the advantages are enormous. While we lose all the perfect-sounding intervals based on small fractions (for example, a fifth is no longer multiplication by exactly $3/2=1.5$, but is now multiplication by $2^{7/12} \approx 1.498$), we can freely change keys with the same notes and all the intervals will sound the same from one key to the next. You gain a lot by giving up on perfection.

Understanding the use of the twelfth root of 2 is vital in building instruments like guitars. The frequency that a plucked string resonates at is inversely proportional to its length (provided you don't change things like the tension or thickness of the string). That is, if you double the length, you cut the frequency in half; if you cut the length into one-third its original length, the frequency becomes three times as large, and so on. Mathematically speaking, this means that if f is the frequency of a string and l is the length of the string, then for some number k that doesn't change (but depends on the type of string used and its tension) you have $f{\times}l=k$. This relationship tells you exactly where to place the frets on a guitar neck. For example, by my measurements, the length of the strings from the nut at the top to the bridge at the bottom (these are the endpoints for the free part of the string that vibrates when plucked) is $l=64$ centimetres. This tells me that the number k is $f{\times}64$, where f is the frequency of the open plucked string. We don't even need to know this. Now the first fret should be placed where the new length of the string, which we'll call l again, will make the frequency up a semitone from f; and as moving up a semitone corresponds to multiplication by the twelfth root of 2, the new frequency will be $2^{1/12} \times f$. So using the formula $f \times l = k$ we get $2^{1/12} \times f \times l = f \times 64$, and with a bit of algebra, we can determine that the length of the string from the first fret down to the bridge must be $l = \frac{64}{2^{1/12}} \approx 60.4$ cm. So the first fret should be 3.6 centimetres from the nut.

SEEING WHAT YOU CAN'T HEAR

The process of decomposing a musical sound into its component frequencies (and their amplitudes) is a mathematical one, called a *Fourier transform*. It is named after the French mathematician Joseph Fourier (1768–1830), who studied such decompositions in his work on heat transfers in physics. This is yet another example of why basic research, that is, research for its own sake, is so critical – you never know where and when down the road it will be needed. Fourier transforms are incredibly useful, cropping up in many, many areas, such as medical imaging. The role Fourier transforms play in music is not often taught at university. I first read about back about them in the late 1980s, and though I didn't make use of the information at the time, I did file it away.

The collection of frequencies of pure tones that make up a complex tone is known as the *spectrum*. Fourier transforms can be used to decompose a sound into its spectrum. Again, our ears and minds are naturally able to do this mathematical computation. The spiral nature of our inner ear's cochlea is critical to decomposing sounds, and enables us, when listening to an orchestra, to pick out the different notes being played.

Knowing about the spectra of notes and how to calculate them with a computer has been very illuminating. For example,

I wanted to find out what makes a voice sound rock 'n' roll. What does that distinctive "gruffy" sound entail? I carried out a little experiment, by plotting and comparing the spectra of two different notes: the sweet opening note that Paul McCartney sings on "Hey Jude," and John Lennon's opening shriek on "Mr. Moonlight" (the song's only redeeming feature).

In looking at the spectrum, I ignored those frequencies that had very low amplitude, as they could have been produced by anything that rattled in the room (like a drum head or a groupie). Mathematical modelling is all about deciding what is important and what isn't.

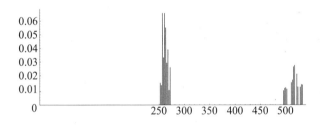

OPENING NOTE OF "HEY JUDE"

In the plot the *frequency spectrum* is shown across the bottom, with the height of a line being the amplitude of the frequency (and so corresponding roughly to the relative volume). What I can see from the spectrum is that Paul McCartney's opening note is made up of a lot of pure tones, but basically there is one loud tone at about 261 Hz (the fundamental, or first harmonic, pretty close to middle C), and another at about 522 Hz. The latter is twice the original frequency, and is the second harmonic. No other harmonics appear at significant volume.

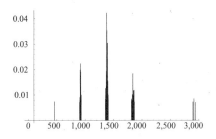

OPENING NOTE OF "MR. MOONLIGHT"

On the other hand, the picture looks much different for John's cry of "Mister . . ." Here I see that there is a small first peak at around 460 Hz (which is almost an A-sharp, though a little flat – more about that soon). The loudest three frequencies are around 920 Hz, 1,380 Hz, and 1,840 Hz. The strongest is around 1,380 Hz, which is the third harmonic to the first tone.

What gives? John manipulates his vocal cavity to emphasize not the fundamental at 460 Hz, but the next three harmonics, the bottom two tones which form a perfect fifth as $1,380 = (3/2) \times 920$. It is this production (with the top of the fifth being louder than the lower note) that seems to give his voice its gruffness. By the way, perfect fifths played at the bottom of the guitar are often called "power chords" for the strength they infuse into a song.

Of course, this is only a preliminary experiment, and I would need to analyze other voices, and perhaps compare different types of vocal productions from the same singer, to make a definitive statement. Knowing a bit of mathematics can open up a whole new world of musical understanding.

COLOUR ME BLUE?

Now more about that A-sharp note that John sings at the beginning of "Mr. Moonlight." The key of the song is F-sharp. An A-sharp, which is the third, is not an unusual note to sing in this key, but it is unusually flat for the key. An F-sharp below A 440 Hz has frequency 370 Hz, so we'd expect the A-sharp, which is four semitones above that, to have a frequency of about 466 Hz. The A-natural would have frequency 440 Hz. The 460 Hz note he sings is somewhere in the middle. Is John just out of tune? (Note for Beatles' devotees: please read on before writing me letters.)

In fact, John is not out of tune, as at this point in the song there are no other instruments to be in tune with. What John probably heard just before singing was not an A or A-sharp, but the tonic chord, that is, the chord that sets the key, F-sharp. And good blues singers (like John) sing not the well-tempered major third, but the *just third*, which has a ratio of 5/4 (one of those nice small fractions). And what is a just third above F-sharp? It has frequency $\frac{5}{4} \times 370 \approx 462.5$, which is pretty much what John nails.

Now you can see why rock and blues singers often sing the flat third when playing the major third; it is an attempt to get a sound close to the "true" blue note. And you will hear blues piano players "crushing" the thirds, that is, playing both the major and minor third at the same time, for the same reason.

I LIKE THE BEATLES, ESPECIALLY THEIR MATHEMATICS

Mathematics plays such a large and varied role in music, far beyond the physics of sound. Having written songs since my early teens, I've noticed that songwriting is a perfect mix of rules and rule breaking. There are chord progressions (sequences of chords) that are well known and well loved, such as the blues progression and the "Heart and Soul" progression. In my band-playing days I had to figure out the music to a new single played on the radio and be able to play it within a night or two. I have done this so often that I can usually decipher the chord changes in a song as it plays, even without a guitar in hand. This process, now that I think of it, is a mixture of practice and deduction; there are chord progressions so common to a genre that my mind naturally sifts through the likeliest possibilities.

This pattern recognition is yet another instance of where I find mathematics in abundance in music. Riffs are often described as a collection of notes or chords that repeat, but I think this definition is lacking. To me, a *riff* is a memorable musical pattern that undergoes some mathematical transformation. Sometimes the transformation is not to change the pattern at all, but merely to move the pattern up or down in pitch, and less often, to reverse it or move its placement in time. The memorability factor depends not only on the catchiness of the pattern, but on how often it appears and how it is transformed.

For example, listen to the Beatles' early songs like "I Feel Fine," "Drive My Car," and "Day Tripper"; you'll find riffs in all of these, and they are probably the parts you most often recognize. You'll find that the riffs, often played by both bass and guitar, are *transposed*, that is, moved up or down in pitch as the song progresses. This act of moving a pattern up or down a set number of positions is really a mathematical one, though practised musicians do it instinctively.

You might think that approaching music so analytically would be a distraction, but I don't find that at all. Mathematics opens up a whole new way to look at riffs, and suggests ways to create them that I might never have thought of otherwise. For example, I was recently listening to Foreigner. If you can dig out your old records (or new CDs) of the group, put on their 1978 album *Double Vision* and listen to "Hot Blooded." You'll hear the rhythm guitar playing a catchy riff, which is based on a short pattern.

The riff is at its core a very simple one, rocking between a chord and its suspended fourth, over and over again. On paper it seems almost too simple to be of interest, but my ears tell me otherwise. Why does the chord riff grab our attention? First, we see that the riff between the first and second bars acts like a *reflection* in a mirror that is set right in the middle line of the staff, along the B line. And in the last two bars of the riff you have a transposition of the basic part of the riff in bar 1.

Now there is more to the riff than just the notes. Rhythmically, the pattern in the second bar is the same as in the first bar but moved back two eighth notes. This sequence of shifts repeats back and forth throughout the rest of the riff.

The net result of all these functions is that a basic musical pattern is morphed into a memorable one. One more Foreigner song comes to mind. Have a listen to "Head Games" from the album of the same name. You'll hear that the chorus "Head games . . ." is on beats 4 and 1. Later on in the song, they deceptively shift the chorus ahead by one beat to place "Head games . . ." on beats 1 and 2. It is such a strong effect that it is almost jarring. Have a listen – what a great title for the song!

Here is something else I noticed that I have never seen mentioned in any analysis of the Beatles' music. The song they conquered America with, "I Want to Hold Your Hand," is completely devoid of any blues notes in the melody, and yet it is a great original rock song, which overflows with energy. But blues notes are almost always required to ramp up the energy of a rock song. So how is it possible? After investigating this question I concluded that what makes the song so memorable are its mathematical tricks!

Let me illustrate part of what I found. The bridge "And when I touch you . . ." moves to an interesting place musically, being a key change (which was a bold move by Lennon and McCartney) up a perfect fourth. Yet it seems to fit the song so well. Why? Well, I noticed that the chords of the chorus, in the key of G, are C → D → G → Em → C → D→ G, and that each chord lasts two beats (except for the final one). The bridge's chord structure goes Dm7 → G → C → Am → Dm7 → G → C, with each chord lasting four beats. But a D minor 7, the first chord of the bridge, has notes D, F, A, and C, and can be looked at as really an F chord with an altered bass note (to a D). So really the chord progression of the bridge is F → G → C → Am → F → G → C, which is exactly *the chord progression of the chorus, moved up by four scale notes*! On top of that, the duration of each of the chords in the bridge is exactly double what it was in the chorus, so there are two different types of mathematical transformations going on, one on the chords, and the second on the durations. It is exactly these mathematical tricks that make the song hang together so well and yet sound fresh every time you listen.

I am fairly certain that none of the Beatles recognized how the chord structure to the bridge was based on that of the chorus (and even more certain that they didn't write it consciously that way). But could they have thought about the transformations subconsciously? I hunted down a copy of their performance of the song on the *Ed Sullivan Show*, and focused in on John Lennon's hands on the bridge. What I found was that he was not playing a D minor 7 at the start of the bridge, but an F chord. This clinched it for me: when he and Paul wrote the song, he must have been thinking at some level of the shifted pattern of chords.

THE ART OF BEING AMBIGUOUS (OR NOT)

Now let's turn to the opening of "I Want to Hold Your Hand"; it is often said that the first few seconds of a song either make it or break it. It takes a while to determine where beat 1, the beginning of a bar, really is, as the musicians enter on the eighth note after beat 3. It is this trick that grabs the listener right off the top.

The trick is a simple (but effective) mathematical one. A more normal transition from the chord C (for two eighths) to D (for a quarter followed by four eighths) would be as follows, starting on beat 4:

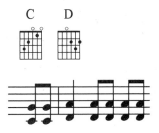

Try playing or tapping out the rhythm – it sounds okay, but pretty boring.

What the Beatles do is shift the pattern back by one eighth note. Mathematically, moving the notes backwards by one eighth note corresponds to using the function $f(x)=x-1$ on the rhythm of the pattern:

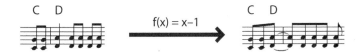

The effect of this simple shift is to transform the beginning from the plodding rhythm above to a brash, captivating one. The whole rhythmic pattern makes finding beat 1 and orienting yourself to the rhythm of the song difficult until bars later. The intentional ambiguity of where beat 1 falls is central to the element of surprise.

In everyday life, it's usually not a great idea to be ambiguous. If your significant other asks if you want to hear the great new song she has written and you say, "I'd like nothing better," you would probably find your slippers neatly parked outside the doghouse before the evening is through. While "double speak" is to be avoided in everyday conversations and books, it is exactly such double meanings that are crucial to a great joke, poem, or lyric. We enjoy the mental game inherent in the mystery. Does John simply light a fire in the fireplace at the end of "Norwegian Wood"? Who is the "Napoleon in rags" in Bob Dylan's "Like a Rolling Stone"? Carly Simon's words to "You're So Vain" are a perfect example of ambiguity: the chorus states "You're so vain, you probably think this song is about you," which, if you are the one she's talking about (and there's a whole mystery there), then of course it is!

Mathematics, you'll be surprised to learn, thrives on such ambiguities. What is less clear is how music can be ambiguous, and what that potential ambiguity can add. In fact, music can have "double entendres" beyond those that fill the lyrics of AC/DC. And underlying these is often a healthy dose of arithmetic.

One mathematical trick used by the best songwriters is something I call "tickling your ears with least common multiples." Musicians have long been attracted to playing patterns of one size against a beat in which the notes are grouped differently. Sometimes patterns of three are grouped; Chuck Berry played many such patterns as he duck-walked across the stage. Here is a typical example of a "tickle," one that I play on guitar. The eight eighth notes in each bar are indicated by x's and the counting underneath. Note how the patterns of three (circled) consist of a quarter note (which consists of two eighth notes) followed by another eighth. The patterns alternately land on a pair of strong beats (with numbers underneath) and a pair of weak beats (with & underneath). The effect is wild and captivating:

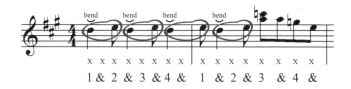

The pattern ends after four repeats, which puts me right on a strong beat, namely beat 3 of the second bar. Why exactly four times? Well, if I want to end off on a strong beat (1, 2, 3, or 4), I need some multiple of 3 (the number of eighth notes in the pattern) to equal a multiple of 2 (the number of eighth notes in a single beat). I could play either two patterns (consisting of three beats), as 6 is both a multiple of 3 and 2, or four patterns (consisting of six beats), as 12 eighths = 6 beats. Playing two patterns is hardly worth my while, but going with four is.

George Harrison often played a similar trick, including throughout "Here Comes the Sun," in the falling lines in the bridge of "Something," and in his famous solo on "A Hard Day's Night." In the latter he played very fast a pattern of three sixteenth notes four times. Again, the least common multiple of 3 and 4 is 12, as 12 is both a multiple of 3 (4×3) and 4 (3×4). Thus the pattern can end neatly after 12 sixteenths, that is, three beats, and it does precisely that. The two Georges (Harrison on guitar, doubled by George Martin on piano) signed, sealed, and delivered the pattern. The key to effectively using patterns of one size against another is to always make sure you have an "exit strategy." That is, know how many patterns you will play so that you leave them in a way that sounds intentional (often, this means leaving the pattern on a strong beat).

Roger McGuinn of the Byrds was a fervent follower of George Harrison, and played a Rickenbacker 12-string electric guitar almost exclusively, due to the influence of George Harrison's guitar playing on the same model in "A Hard Day's Night." When I listen to some of the famous guitar riffs of the Byrds, what stands out is that Roger not only shared George's taste in guitars, but his taste in mathematical tricks as well! In both "Turn, Turn, Turn" and "Mr. Tambourine Man," Roger played groupings of three eighth notes on his 12-string guitar for the catchy riffs. It seems they shared a similar intelligent mathematical mindset and were both intrigued by the use of one rhythm against another. I think Jerry Garcia, the late great lead guitarist of the Grateful Dead, summed it up best when he said, in the December 2005 issue of *Guitar Player*, "If the band is playing in 7/4 time, I might play in 4/4. Doing that, you begin to notice certain ways the two rhythms synchronize over a long period of time."

You can hear other examples of the "least common multiple tickle" in Eddie Van Halen's opening to "Panama" (it starts one eighth beat after beat 1, and hence is even more deceptive), and Angus Young's work in "Back in Black." In the latter, not only can you hear it in the opening riff but also in the chords behind the chorus. Sometimes you can hear good session players pass the pattern back and forth among themselves. Pick up a copy of Lee Ann Womack's "I Hope You Dance." In addition to being a lovely, inspirational country song, it's a great example of the least common multiple tickle. Listen to the background musicians. They trade 3-against-4 patterns amongst themselves throughout the song.

It's unlikely that any of these musicians would consider themselves mathematicians. In fact, George Harrison once remarked that he never did well in mathematics in school. But musicians often use mathematics unconsciously, and in that sense they are truly mathematical.

IT'S BEEN A HARD DAY'S NIGHT

My mathematical excursions into pop music came to a head in 2004 – the fortieth anniversary of the Beatles' first (and best) movie, *A Hard Day's Night*. This brought me back to the title song, whose opening chord is etched in my mind. Movie reviews of *A Hard Day's Night* always seem to make mention of it; Roger Ebert wrote, "Perhaps this was the movie that sounded the first note of the new decade – the opening chord on George Harrison's new 12-string guitar." In "Hey, Ho! Let's Go! The Greatest Intros in Rock and Roll" (*Palm Beach Post*, July 3, 2004), this introduction stood at number one, with the following comment: "The guitar chord heard 'round the world.' Musicologists still debate what the chord is, where exactly George Harrison fingered the frets on his 12-string guitar. Whatever. That 'Bwwwaaaaang' rocks." To wax a bit poetic, I imagine myself taking a dive off a dock on a sweltering hot day. The opening chord of "A Hard Day's Night" is the sound I think I would hear as my body cuts through the cool, sparkling water of the lake.

The importance of the opening chord was apparent to the Beatles back when the song was recorded. In *The Complete Beatles Recording Sessions*, author Mark Lewisohn quotes the Beatles' producer, George Martin, as saying, "We knew it would open both the film and the soundtrack LP, so we wanted a particularly strong and effective beginning. The strident guitar chord was the perfect launch."

The Chord is central to rock 'n' roll history, but I have yet to hear of a guitarist able to play it in all its majesty. The question of how George Harrison played this opening chord is a frequent question on Beatles-related websites. Alan W. Pollack, who published online in-depth musical analyses of all of the Beatles' recordings, states: "I've seen better people than myself argue (and in public, no less) about the exact guitar voicing of this chord and I'll stay out of that question" – though he does go on to say what notes he hears (D, F, A, C, and G).

Most guitarists have their favourite approximation of what George Harrison played. For example, you can browse transcriptions to see the following (note that guitar parts are usually scored an octave up from where they sound):

VERSION 1: G C F B-flat D G (a favourite because of its ease of play; just play a barre chord at the third fret).

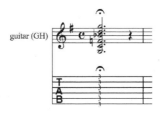

VERSION 2: G D F C D G (another favourite, and one that often appears on Internet sites as the "one" that George played).

VERSION 3: This one has George, John, and Paul playing, with George playing G D G C D G, slightly different than in version 2 (on his 12-string electric), while John plays D G C G and Paul plays a D on his bass (this is the transcription from *The Beatles – The Complete Scores*).

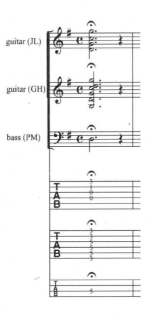

Why some 40 years on is there still a mystery surrounding how this chord was played? Certainly, there are only so many ways George could have placed his fingers on the fretboard, right? My ear is pretty well tuned to pick out notes from a chord, but here I had no luck. Is there any way to tell which of these three versions is the correct one? Certainly many musicians and musicologists have tried to penetrate the mystery of The Chord, and I could have relegated the question to the "interesting but unsolvable" trash heap next to my desk. But I find it harder and harder to separate out my mathematical life from my non-mathematical one, so I decided to see what mathematics might say about the chord.

When a sound is digitized for a CD, the amplitude is recorded 44,100 times every second (so your beautiful-sounding CDs are nothing more than a long collection of numbers). This conversion of sound to numbers is good enough to produce very accurate, high-quality renditions of the original.music. In fact, sampling the sound 44,100 times per second ensures that all of the frequencies humans can hear will be accurately recorded and reproduced.

Yet what is hidden in the process of recording music are the individual notes that were played, and how they were played; all that remains are the millions of numbers listing the combined amplitudes of the sounds at each 1/44,100th of a second. Is there any way to reconstruct what the pure tones are that make up the music?

This is where Fourier transforms come to the rescue. All you need to know for the rest of this story is that the Fourier transform is a mathematical operation that allows us to reconstruct the original frequencies and amplitudes from the individual list of numbers of the digitized sound (there is a variety of software that can carry out the calculations). The important thing is that the list of numbers has to come from a sound that has the same notes playing (at roughly the same loudness) throughout the selection, so the Fourier transform is not useful for trying to map out all the notes that are heard throughout a song, as most often different instruments are playing different notes at different times, overlapping one another.

But, hey, we are in luck: in listening to the opening chord in "A Hard Day's Night," we hear that no new notes are struck in the middle; one chord is played and hung out to dry. So what I did was take the CD, transfer the digitized song into a sound-editing program on my Mac, and select a segment of about one second in the middle of The Chord (chosen so that the sound wouldn't vary in volume much over the segment). I got a huge list of numbers (58,752 to be exact). I saved the sound as a file on my computer and ran the Fourier transform on the list of data.

Well, even for what must be a fairly simple sound (after all, how many notes could be ringing on George's guitar?) I found out that there were 29,375 frequencies present! But I realized that what was included was not only the notes being struck, but also harmonics, as well as any other frequencies that might come from some shake, rattle, or roll in the studio during the recording of this chord.

WE *CAN* WORK IT OUT

I was after the loudest notes, as these correspond to the funda-mental notes being struck (though there will probably be some of the louder harmonics present, along with possibly some other loud rattles). I therefore removed frequencies of lower amplitude, and played the chord resulting from what remained, making sure that the sound seemed faithful to the original, perhaps going a little overboard and admitting some frequencies that weren't fun-damentals into consideration. Here is what I found (the first number in each list is the frequency, the second its amplitude, which shows its relative volume):

Freq. (Hz)	Ampl.	Freq. (Hz)	Ampl.
110.34	0.0600967	299.494	0.0298296
145.619	0.025485	392.57	0.0309716
148.621	0.0264278	438.358	0.0286329
149.372	0.0656018	524.678	0.0680974
150.123	0.175149	587.73	0.020613
174.142	0.0275547	588.48	0.0310337
174.893	0.0380282	589.231	0.0231753
175.643	0.0407103	785.141	0.0323532
195.159	0.0405164	786.642	0.0251928
218.428	0.0448308	787.393	0.0268553
261.964	0.0302402	960.784	0.0228509
262.714	0.0234502	981.801	0.02242

Freq. (Hz)	Ampl.	Freq. (Hz)	Ampl.
1050.86	0.0687151	2368.93	0.0221358
1185.97	0.0372155	2371.19	0.0212846
1286.55	0.0231789	2371.94	0.0436633
1314.32	0.03819	2372.69	0.036042
1320.33	0.0223535	2637.65	0.0261839
1321.08	0.0494908	2638.4	0.0237794
1488.47	0.0241328	2754.	0.020001
1632.58	0.0205742	2763.76	0.0493617
1750.43	0.0234704	3083.52	0.0332062
2359.93	0.0366079	3147.32	0.0293723
2367.43	0.0267098	3148.07	0.0418507
2368.18	0.0755327	3158.58	0.0285631

As you can see, there are 48 frequencies at this threshold of loudness, ranging from about 110 Hz (cycles per second) to 3,158 Hz. Not all of these are likely to be notes that were played, but some are. My work was to figure out which were which.

Now frequencies are just fine, but musicians play notes. I needed a reference note, so I picked A 220 Hz (the A below middle C) and converted the frequencies to the number of semitones above or below A 220. Here is the list I got:

−11.9466, −7.14367, −6.79035, −6.70313, −6.61635,

−4.04686, −3.97239, −3.89825, −2.07421, −0.124124,

3.02237, 3.07191, 5.34031, 10.0254, 11.9353, 15.0472,

17.0118, 17.0339, 17.056, 22.0254, 22.0584, 22.075,
25.5205, 25.8951, 27.0719, 29.1659, 30.5752, 30.9449,
31.0238, 31.0337, 33.099, 34.699, 35.9056, 41.078,
41.133, 41.1385, 41.1439, 41.1604, 41.1659, 41.1714,
43.0042, 43.0091, 43.7514, 43.8127, 45.708, 46.0626,
46.0667, 46.1244

Now wait a minute, you might say. Shouldn't all the numbers be whole numbers? After all, every note is some whole number of semitones away from A 220, not fractional parts of a semitone. Well, in a perfectly tuned world, that would be true. But instruments are never perfectly in tune. We see in particular that some of the notes were a fair bit out of tune; for example, the fifth number, –6.61635, is probably a note that should be seven semitones down from A (that is, a D below A 220). To handle the fact of imperfect tuning of the Beatles' instruments, let's round each number to the closest whole semitone. In musical circles, middle C is written as C4, with the second number indicating the octave, so A 220 Hz is written as A3. Here are the frequencies above rounded to the nearest semitones:

A2, D3, D3, D3, D3, F3, F3, F3, G3, A3, C4, C4, D4,
G4, A4, C5, D5, D5, D5, G5, G5, G5, B5, B5, C6, D6,
E6, E6, E6, E6, F#6, G#6, A6, D7, D7, D7, D7, D7, D7,
D7, E7, E7, F7, F7, G7, G7, G7, G7

Most of the notes come as no surprise; many of them appear in the various versions of The Chord and the notes confirm that the spooky mathematics introduced does match well with what very musical ears have been hearing all along. But what I want to do now is argue about *what* notes were played and *how* they were played, and for this we'll resort to deductive reasoning.

Some of the notes (especially in the higher range) must be harmonics, as they are well beyond what instruments can play. In fact, the range of a guitar is from E2 to about E6 and a bass guitar from E1 to about D4 – so the notes could have arisen on either George's or John's guitar or on Paul's bass.

We see now why the three well-known transcriptions of the opening chord must all be wrong – each has a low G2 being played, but this note is definitely missing! There are other problems with the transcriptions as well. We could leave it at that, but I'd like to explain the rest of the story.

George's wonderfully twangy, brand new Rickenbacker 360/12 twelve-string guitar can definitely be heard on the solo in "A Hard Day's Night." Thus it seems safe to assume that George used this guitar on the opening chord as well. The 12-string guitar has each string doubled, with the bottom four in octaves, so the strings are, from lowest to highest, E2 E3 A2 A3 D3 D4 G3 G4 B3 B3 E4 E4. It seems reasonable that notes on strings of roughly the same thickness struck on one instrument would be roughly of the same amplitude. Looking back at the frequencies and their amplitudes, we see that one D3 is extra loud, with an amplitude of 0.175. I take this as a bass note from Paul's Hofner bass (no other single frequency is nearly as loud).

The A2 and A3 can also be paired off together. I would suppose they came from George's 12-string (a nice open pair of strings). But there's a problem. Even with one of the D3s accounted for on Paul's bass, what about the other three D3s?

Only one can come from George's 12-string, and even if John played another one of them on his six-string, there is still a third to account for! There is no evidence that any guitar was multi-tracked, at least not on this opening chord. The three F3s create an even bigger problem. For no matter how George plays an F3 on his 12-string, an F4 should be heard as well, and there is no F4 at all present! What is going on?

BY GEORGE, I THINK I'VE GOT IT!

It took a while to figure this one out, and I needed to throw out a deeply held conviction. It came to me that the Beatles' record producer, George Martin, is known to have doubled on piano George Harrison's solo on the track. Could the chord that I and so many others have attributed to George Harrison have another George in the mix? Could the most famous rock 'n' roll guitar chord be part piano? I researched the construction of pianos. It seems that pianos have three strings for every note; a hammer strikes all three at the same time to produce a sound. That solved the problem of the three F3s – all could have come from a piano playing F3! The fact that the frequencies of the three F3s were slightly different only confirmed my deduction, for each string on a piano is individually tuned, and is likely to be slightly off from the others in the "triple."

But what about the three leftover D3s? If all belonged to a single piano note, then where would the single D4 come from? Not from George Harrison's 12-string (as a D3 or another D4 would be present) and not from George Martin's piano (as otherwise there would be three D4s). I did a bit more research and found that the bottom 10 pitches on a piano are single strings that change to pairs of strings, and around C3 they change to the usual triples of strings. Breaking at C3 was still a problem, but after sticking my head into a few grand pianos during a quick trip to the local piano store I confirmed that indeed there are some grand pianos (those of medium-length) for which the break occurs right after D3. Success! This told me that two of the D3s were played on the piano, as well as revealing something about the size of the piano in the Abbey Road studio!

These deductions led me to decide that what George Harrison played on his 12-string was nothing like any of the transcriptions: he played A2 A3 D3 D4 G3 G4 C4 C4, most likely on string sets 2 through 5 – eight strings with six open strings in total (for a great chiming effect). George Martin played D3 F3 D5 G5 E6 on the piano. The other notes are fairly high, and could be attributed to harmonics of these notes, except that there is a loud C5, which could have been played by John, high up on his six-string. There is also one extra E6 that is unaccounted for, which I take as a harmonic.

The response to the publication of my mathematical research into the opening chord of "A Hard Day's Night" was much more than I ever could have imagined. Inside of one week, I made front-page news in the *National Post*, I was interviewed on the CBC, and over subsequent months I had two articles published in *Guitar Player* magazine, where I can safely claim that those articles contained and will likely contain more mathematics than any guitar player is likely to read in any other column, *ever*. My 15 minutes of fame. On the other hand, I also had Beatles fanatics e-mailing me, telling me that I was desecrating the Beatles by trying to uncover any mysteries surrounding them.

Initially I was a bit disappointed with what I had found out about the song. I started learning the guitar the day I heard my first Beatles album, and I have always associated the Beatles with the guitar. Now every time I play "A Hard Day's Night," I can't help but hear the piano in the mix (try it!). But mathematics takes me where it takes me – I can only follow.

I learned things about the music that I wouldn't have otherwise known. First of all, George Harrison plays three consecutive fourths (actually, a total of six, counting the eight strings played) on the guitar (A → D → G → C). Six of the strings are "open," giving a bright chiming sound. The use of fourths in rock 'n' roll was (and still is) unusual; fifths play the dominant role in rock. So on that basis The Chord is already rather unique.

Moreover, we spoke about *combination tones* earlier, notes that are not played but are perceived when we hear certain pairs of tones together. In particular, when we hear a fourth, we perceive an extra tone that is a fifth below the bottom note. This I believe accounts for why the bass note in Version 3 (from *The Beatles – The Complete Scores*) is scored an octave lower than what it really is. In any event, the choice of fourths adds a "heavier" feel.

The notes played on the piano interleave well with the notes on the 12-string, starting a bit higher (at D3) from the lowest note played on the guitar and ending higher (at E6). The amplitudes show why the piano is so well hidden; it is mixed perfectly, with amplitudes almost identical to those of the higher strings played on Harrison's guitar. One crucial new note is added in on the piano, an F; this is the root of the one out-of-key chord (namely, F) in the piece (and the one that helps form the ending of the song as well).

Do you notice how the opening chord has a "wah wah" effect? This is, I believe, a fortuitous consequence of the fact that not all of the instruments were as in tune as they could have been. Two tones that are close in frequency but not identical are perceived as "beats," with a "wah wah" effect. This is nothing more than our brain hearing a mathematical identity, just like the ones you were forced to prove in high school. Whenever you combine two sine curves with close frequencies, the result is a curve whose frequency is right in the middle, but that has its volume swell and fade continually; you don't perceive the sound as two distinct notes. Among others, there is a 5 Hz difference between the D3 George Harrison plays and the other D3s played by Paul and George Martin. It's interesting to note that George Martin says in his book *All You Need Is Ears*, "I think that being out of tune, provided it's tuneful, is in itself an attraction." Case in point!

What accounts for the large number of frequencies (about 29,000) present in The Chord? It's likely that the piano notes were struck with the sustain pedal down, allowing for a full sound as other notes (and their harmonics) rang ever so softly as well – a wonderfully subtle addition to the wall of sound.

Finally, what about the opening "strident guitar chord" that George Martin spoke about? In *All You Need Is Ears*, Martin makes a point of saying that "it shouldn't be expected that people are necessarily doing what they appear to be doing on records" and likens recording to filmmaking, where all sorts of effects are carried out in the background in order to create illusions. The opening to "A Hard Day's Night" may no longer be the greatest *guitar* chord in rock 'n' roll, but it still is the greatest chord in rock 'n' roll.

As I take out my acoustic guitar to play for our guests, I think of
Tom Lehrer, a mathematician well known in the 1950s and 1960s for
singing parodies whose lyrics ranged from funny to outrageous.
Like Tom Lehrer, I chose the path of mathematician over that
of musician, precisely because mathematics is at least as much fun
(and the safer-paying gig). But tonight I tune up my guitar anyway.
Like mathematical research, you can't keep a good set of lyrics down.

13

TO COUNT, PERCHANCE TO SLEEP . . .

It's late in the evening. I don't usually go to bed before midnight – a holdover from my university days. Out of the kitchen window, I can see a clear night sky, filled with stars. Are there infinitely many? I'll never know for sure, but I think so, as the alternative is a point at which space is empty of stars forevermore, and I find that even harder to imagine.

Constellations start to take shape –chance arrangements that are part of the beauty. I guess we never tire of finding patterns. I make a small snack, sit down at the kitchen table, and ponder the grander scheme of things. And yes, mathematics too.

Even non-mathematicians would have to agree that our world is a mathematical one. Man used mathematical principles to explain why reality is as it is. Newton discovered rules that governed motion, and Einstein built upon them, refined them, but these scientific laws were expressed in mathematics neverthe-less. The principles of genetics depend on probability. We've seen mathematics lurking about when we sit down to eat, when we hop on the treadmill to exercise, when we stroll outside to appreciate nature, when we make decisions, and when we listen to great music. The question isn't really where math is, but where *isn't* it?

A world in which mathematics rules supreme is one of the greatest mysteries of all. Why are the principles on which so much of our world is based inherently mathematical? Humans have an extraordinary ability to understand, utilize, and create mathematics.

In fact, there have been experiments that show that babies as young as five months old have the innate ability not only to distinguish small numbers, but also to do simple arithmetic! How did the experimenters determine this? They showed infants one puppet on a stage. Then a screen popped up and hid the puppet from view, while a second puppet appeared on stage, moving from in front of the screen to behind it. When the screen disappeared again, there were either two puppets (as expected), or a single puppet (a surprise, and incorrect mathematically, as 1 puppet plus 1 puppet should equal 2 puppets).

The experimenters found that the babies didn't stare for very long if the expected number of puppets, 2, was present, but looked for longer if only one was there. The level of surprise in the infants when faced with the wrong arithmetic consistently indicated that even very young children can do very basic math-ematics. (I have tried the puppet trick with my friends and have always found a striking level of surprise . . .)

So it seems that from a very early age we all have some natural mathematical ability. Wouldn't it be a shame to let it go to waste?

ASK NOT WHAT YOU CAN DO FOR MATHEMATICS . . .

A little mathematics can take us far. And it's not only the calculations and concepts that can be so helpful but even the perspective mathematics offers. Life is full of problems, begging for solutions. We might have an approach or two that we often use, probably with mixed results. What mathematics can do is add to our bag of tricks for problem solving, thereby dramatically increasing our chances of success. Mathematics can sweep away some of the limits we impose on ourselves, if only we are receptive. One of my favourite quotes is Paul Erdös' "My brain is open," a sentiment I think we should always aspire to.

Erdös had a knack for finding ground-breaking creative approaches to old mathematical problems. Sometimes when Erdös wanted to know whether some particular graph existed, instead of trying to build it, he would take many points and toss in lines between the points at random. He could often show that at least one of the resulting graphs would have just the property he was looking for.

Here is an example. You might want to ask if you can find a graph so that whenever you take *any* two, non-overlapping collections of five points, you can always find another point that is joined to all of the points in the first collection and none in the second. By using Erdös' technique we can show that there is a graph with exactly this property. In fact, almost every one of the resulting graphs would have this property, even though we would be hard-pressed to find a single example! Usually, to believe something is true you have to see it or build it, but Erdös showed that sometimes you don't need to do either.

The problems we find ourselves dealing with on a daily basis are likely less abstract than most mathematical problems, but you can appropriate some useful approaches nonetheless. Are you stuck in a situation, a puzzle with seemingly no way out? What about dividing the problem into smaller, manageable sub-problems, and working on each one? Or how about working backwards, from where you want to be, instead of worrying about how to get there? Is there some symmetry present, that is, some parts of the problem that are somewhat like one another, and can you use this to weed out your possibilities? Or is it worth gathering some more data, and searching out useful patterns? We have seen all of these techniques in this book, and thinking about them explicitly may be the first way out of the box.

And don't we often hear that employers are always looking for people who think outside the box. A recent article on the *Wall Street Journal's* website looked at a study of the 100 best and worst jobs in the United States, rated on a variety of scales (working conditions, pay rates, and so on). What do you think was rated the best job? Mathematician! Mathematical knowledge is something you can be certain will always be needed, and will never go out of style.

Mathematics can improve your life even outside the workplace. Who wouldn't like to be more creative? I know I would. When I'm writing music and I get stuck, one of the first things I do is run through some approaches I've learned in mathematics. I transform a pattern of notes I've written; I flip some notes around; I work back from the song's climax to build up to it; or I add in a few chords or notes at random to spice things up. The result may not be good, but on the other hand it might be unusual, exceptional even. Don't worry if an idea doesn't work out; the main thing is to keep having those ideas and to be unafraid of trying them out.

On a recent trip to Nashville, I stopped by the world-renowned recording studio Blackbird Studio. The designer of the famous Studio C there, George Massenburg, has a strong mathematical outlook. One of the biggest challenges in constructing a recording studio is getting the acoustics right, as echoes can greatly affect the end product. What Massenburg did was cover the walls of the studio with various lengths of wooden poles, and used number theory to decide the length of each rod. The outcome was what looks like a random arrangement of wood that made possible the lushest playback of the Beatles singing "Because" I had ever heard. It was otherworldly; I didn't know whether I was more impressed by the sound or the use of mathematics. It took a lot of guts to spend a fortune on a mathematical idea – a new and untried one at that – but great rewards awaited Massenburg (and the artists he recorded).

IS THE GLASS HALF-EMPTY OR HALF-FULL?

The mathematical construction of the recording studio brings to mind an apparent contradiction, that of using a defined mathematical procedure to approximate randomness. Chaos theory is based on the fact that the result of many repetitions of the same old process can produce unpredictable results. That's why weather prediction is so difficult – the short-term predictions are usually pretty accurate, but the long-term ones are only as good as a shot in the dark.

Mathematicians have always been attracted to paradox – the seeming inconsistency is an enticing puzzle that demands to be solved. A classic paradoxical statement is "This sentence is false." If the sentence is true, then it is false, and if it is false, then it is true. If you watched the original *Star Trek,* you probably remember the episode where Captain Kirk short-circuits a race of robots by telling them a variant of this paradox.

One of the earliest paradoxes that I ever heard is still a favourite. Suppose you are a librarian, and in your library, you decide to create a new master book called, naturally, "Master Book," to be placed on a shelf with the other books in the library. Inside this book you list the name of every book in the library that doesn't mention its own title within its pages (on the cover doesn't count). So for example, if I include the title *Our Days Are Numbered* here, you wouldn't list *Our Days Are Numbered* in the Master Book.

Now here is the million-dollar question – should you include "Master Book" in the master book? Think about it. You should be able to convince yourself that if you don't include it in the book, you should, and if you do include it, you shouldn't. Time to get a different job.

The librarian paradox is an example of a self-referential statement. Statements that are self-referential are prone to paradox, and untangling them often involves taking a viewpoint that is up one level from what you are talking about. There is, for example, a level of mathematics where we do mathematics. But there is a higher level, a *meta* level, where we talk about the process of proving things in mathematics. And, of course, there may be a *meta meta* level where we discuss how we go about discussing how to prove things in mathematics. It gets rather mind-bending.

There was a whole movement in mathematics around the turn of the last century to try to "axiomatize" mathematics, that is, to find some rules, or axioms, from which everything true in mathematics could be proven. This whole approach came crashing down when the Austrian-American mathematician Kurt Gödel proved that no matter what axioms you tried to take, there would always be true statements about numbers that you couldn't possibly prove.

Mathematics has the last laugh – even on itself – proving that there are always true things that are unprovable. Mathematics may not be the answer to everything. Still, it can take us on a lovely ride.

That's enough philosophizing. Too much thinking can't be good, at least not this late at night. I'm heading off to bed. You're welcome to stay and think about a problem or two. When you're done, please turn out the lights.

ACKNOWLEDGEMENTS

This book has been a labour of love, a long, induced labour, and finally I can see the head crowning. My wife feels it's entirely my turn this time round. There are so many people I'd like to acknowledge, so I thought, of course, I'd create an ordered list. Starting at number 10 . . .

10. To my many students over the years: I've learned a lot from all of you.

9. To my colleagues in the mathematical community: To Richard Nowakowski, Derek Corneil, Eric Mendelsohn, Ivan Rival, Charlie Colbourn, Vojta Rödl, and the countless (well, almost countless) others who have made doing mathematics so enjoyable.

8. To Paul Erdös: he showed me not only what mathematics could be, but what a mathematician could be.

7. To the Beatles: I have learned many a lesson on research and life from their music. Now that there's been a Pope John Paul, don't you think it's time for a Pope George Ringo?

6. To my agents, Shaun Bradley and Don Sedgwick, who encouraged me to write this book, and to the folks at my publisher, McClelland & Stewart, and especially my editor, Jenny Bradshaw. Jenny got me – the math and jokes. And I simply could not ask for more.

5. To my friend Cynthia Martin, who read and edited my original article on mathematics and the Beatles, where all of this started. Her generous advice and incisive suggestions have been much appreciated.

4. To my siblings – Arthur, Renie, Paul, and Milton. I have had the good fortune of having you not only as brothers and sister, but also as friends.

3. To my sons, Shael and Zane. As Larry said of his brothers, Darryl and Darryl, on the *Newhart* show, "Sometimes that DNA goes so right."

2. To my parents, Reuben and Ester. Being born is pretty much a lottery and, boy, did I win the big one.

1. And most importantly, to my wife, Sondra, the reason I get up the morning (she won't let me sleep in). I simply couldn't have written this book without her support and help; she read every word, made suggestions, and laughed aloud when editing out some of my more risqué jokes. To my wife, my girlfriend (still, after 17 years), and my best friend – you prove that even a mathematician can be lucky enough to pull a pretty bird.

INDEX